The Art of Being a Scientist

This is a hands-on guide for graduate students and other young researchers wishing to perfect the practical skills that are needed for a successful career in research. By teaching junior scientists to develop effective research habits, the book helps make the experience of graduate study a more efficient, effective, and rewarding one. Many graduate students learn these skills "on the job," often by doing them poorly at first, with the result that much valuable time can be lost: this book will help prevent that. The authors have taught a graduate course on the topics covered in this book for many years, and provide a sample curriculum for instructors in graduate schools who wish to teach a similar course. Topics covered include:

- how to choose your research topic, department, and adviser
- how to make a workplan
- the ethics of research
- how to use the scientific literature
- how to perfect your oral and written communication
- how best to publish papers
- how to write proposals
- how to manage your time effectively
- how to plan your scientific career and apply for jobs in research and industry

The wealth of advice offered in this book is invaluable to students, junior researchers, and their mentors in all fields of science, engineering, and the humanities.

The Art of Being a Scientist

A Guide for Graduate Students and their Mentors

ROEL SNIEDER AND KEN LARNER

Department of Geophysics and Center for Wave Phenomena
Colorado School of Mines, Golden, CO 80401, USA

CAMBRIDGE
UNIVERSITY PRESS

CAMBRIDGE UNIVERSITY PRESS
Cambridge, New York, Melbourne, Madrid, Cape Town, Singapore, São Paulo, Delhi

Cambridge University Press
The Edinburgh Building, Cambridge CB2 8RU, UK

Published in the United States of America by Cambridge University Press, New York

www.cambridge.org
Information on this title: www.cambridge.org/9780521743525

First published 2009

Printed in the United Kingdom at the University Press, Cambridge

A catalog record for this publication is available from the British Library

ISBN 978-0-521-74352-5 paperback

In memory of Rodney Calvert, a gentleman whose inquisitive nature, contagious enthusiasm, and caring attitude manifested the art of science in both his personal and professional life.

Contents

1 Introduction

You're planning to pursue graduate education or perhaps are in an early stage of graduate study in science or engineering, or perhaps the humanities. You might therefore be thinking that the future course of your graduate studies and career thereafter are well set in place: you can now proceed with your course work and research largely on automatic pilot. The experience of most graduate students, however, is unfortunately to the contrary. While you know not to expect smooth sailing in your studies and research, you might be unaware that many roadblocks, sources of frustration and angst, and much wasted time during and after graduate study can be avoided or at least substantially minimized, perhaps making the entire experience largely satisfying, indeed joyful – one in which you thrive.

This book is a practical guide with two primary goals. The first is to help make the experience of graduate study for students early in their graduate program in science, and for senior-level undergraduates intent on entering such a program, be an efficient, effective, and generally positive one. The second goal, consistent with the first, is to help those students and other junior researchers develop effective research habits.

While some will choose to read this book from cover to cover, many will find benefit from reading selected chapters in depth at different stages of their university or professional careers, perhaps returning to specific chapters as needed.

For three reasons, the contents of this book are not focused on the disciplinary aspects of research. First, one cannot hope to cover all research fields in a single book nor would the reader be drawn to the intricacies and subtleties in specific research areas across the breadth of scientific disciplines. Second, the authors of course do not have close to the necessary disciplinary expertise for the many different fields in science. Third, it would defeat the purpose of the book, which aims to offer broad counsel for aspiring researchers, independent of discipline. The book therefore focuses on practical aspects of research that are of relevance for students and young researchers in the sciences in general. Throughout the book, we use the word *science*, but since many aspects of research are generic for both science and engineering, the guidance offered here should be of comparable relevance as well to engineers intent on a career in or closely related to research. Research approaches differ among various fields, but much of the presented material should also be of benefit in the graduate and early research careers of students in the humanities and medicine, as well.

It's a good bet that none of Isaac Newton, Francis Crick, nor Charles Darwin would have been interested in or had need for this book. They had an innate sense of how to blaze profound trails in science, driven by their curiosity and aided by clear focus of attention. The vast number of budding scientists at early stages of their graduate-school careers, however, could benefit from many of the suggestions in the book. Decades back, when we two authors were students, we would have done well to have such guidance.

Certainly Ph.D. and M.Sc. programs can be successfully completed without access to the suggestions offered in this book. Countless scientists have made it through graduate school without such guidance en route to their often outstanding careers, so the graduate experience can and does work. Ultimately getting to the stage of writing and defending the thesis undoubtedly constitutes one level of successful completion. A central tenet of this book, however, is that a more satisfying degree of success would be the defending of research that is an excellent fit for the candidate in the sense of its match for his or her talents and interests, and getting to that moment efficiently, with a minimum of time, pain, and consternation, and a maximum of enjoyment and professional success.

For any of a number of reasons, you have willingly chosen – likely been drawn to – a career in science, and this choice is one that you probably have made with some excitement. Many of you can be said to love the ideas in science and technology that have attracted you. It is all the better therefore that your pursuit of science, from graduate school through your subsequent career, turns out to be a joyfully memorable experience.

Why this book? Unquestionably, the pursuit of a Ph.D. or Masters degree can be a challenge at times. Much needs to be learned along the way – about both the science and the scientific technique – and that learning takes time, discipline, and hard work. The suggestions offered in this book won't remove that essential challenge. Rather, what they aim to do is to make the student aware of pitfalls that often are the source of unnecessarily painful experiences, damaging self-doubt, and needless prolongation of the time for completion of the graduate program.

To a large extent, the endeavor of graduate study can be characterized as one of learning on the job. In a way it resembles the medieval method wherein an apprentice learned a trade by working with a master. Most graduate students are educated in much the same way; by working with an academic advisor, they learn the trade of science while doing research. The process of learning is largely based on imitating the adviser and other faculty members, and being corrected when things don't go smoothly.

This on-the-job-training method of education works well enough in that most graduate students will have learned to carry out science independently by the time they receive their degree. This, however, does not mean that it offers the optimal way to prepare graduate students for a career in science. This mode of learning has two disadvantages. First, learning on the job implies that making mistakes often plays an unnecessarily large role in the learning process. Mistakes, to be sure, can be a powerful and valuable teacher. Those that could have been readily foreseen and avoided but were not, however, cause delays and are potentially harmful stumbling blocks. Second, many of the lessons for junior researchers can be conveyed more effectively by exposing them explicitly, as in this book, rather than implicitly, as when all learning is on the job.

Consider the following example. Writing is difficult for most people. It certainly can be so for a beginning graduate student who starts to write her first major paper. Often the student agonizes and writes the first draft with great difficulty. She next gives this draft to her adviser. More often than not it can be expected that the adviser will disagree with aspects of the basic structure of the manuscript, the conclusions that are drawn, or the style in which the results are presented. The resulting necessity of having to do a major rewrite of the manuscript then contributes to mounting frustration and often loss of confidence of the student. As shown in Section 10.3, this problem largely can be avoided by developing effective writing habits and by being mentally and emotionally prepared to understand that iterations of the drafts between adviser and student constitute an expected and potentially productive process. A good deal of important scientific advance takes place during the draft-iteration process.

As in learning any new field or occupation, mistakes along the way during graduate study are inevitable. With the benefit of lessons learned from the experiences of others, however, many *unnecessary* mistakes and pitfalls, while engaged in a research project, can be avoided, thereby helping to shorten the time spent in graduate study and make that study a more positive experience. Since most graduate-student projects are carried out under both time-pressure and a tight financial budget for the student, avoidable delays are undesirably costly. Of more concern, the related frustrations for the student lead to the second disadvantage of this mode of learning – a most unfortunate one: the drop-out rate of students from graduate school is higher than it needs to be.

Graduate students can be better prepared for a career in science if they have an explicit understanding of the many issues that arise as they are about to embark on that career. To help prepare students for a career in science, one of us started the course "The Art of Science" at the Colorado School of Mines. The course is aimed not at providing a philosophical treatment of the process and nature of science. Instead, it aims to help prepare graduate students for a career in science by teaching them practical skills and offering insights that can be beneficial in carrying out science, with a primary goal that their graduate-education experience be both enjoyable and efficient. Many of our colleagues at different institutions agree with our perception that there is a real need in most graduate programs for education along the lines of this course. It provides beginning graduate students with general skills to carry out research more effectively at an early stage of their career. The following quotes taken from evaluations of the course are fully representative of benefits that students perceive:

> "We are getting counseling that students typically don't receive from an adviser. I for one appreciate it."
>
> "I enjoyed your approach to research planning, problem solving, and personal development. I will certainly recommend your class to other students!"
>
> "I have learned that there are many people encountering the same kinds of problems that I have in my research."
>
> "The course was very useful, although I think it would have been even more useful if I had taken it when I just started my Ph.D. research."
>
> "This class should be a must for any graduate student."

These statements provide support that a course that covers material such as that in this book can be highly useful for graduate students. Students consistently indicate that they feel such a course would be most effective if taken early in the graduate-school career. It is a relief for students to discover that many fellow students struggle with similar issues, and that they can often help and support each other in resolving those issues.

Offering this course as a 1-credit course for one semester has been the best compromise between the depth and breadth of material covered, and the needed time-investment by students. Appendix B shows a possible curriculum for such a course that serves to give instructors ideas about topics and activities that one could use.

Material in this book can also be selected for use in a short course with a duration anywhere from a few hours to several days.

For whom is this book written? This book is written primarily for students who pursue a Master's or Ph.D.-degree in science, engineering, or the humanities. Since we describe the development of effective research habits, the material is also of relevance for undergraduate students who are planning a career in research, or who simply want to know what it takes to be an effective researcher and what a career in research might involve. Although the advice in this book is primarily aimed at students, this material can also be of relevance for advisers of students as it could help them to be more effective in their role as mentor. Last, this book provides ideas about the practice of research that could be of value to freshly degreed researchers, as well as to those already well into a research career.

An overview. Numerous excellent textbooks on the philosophy of science have been written. Rather than giving an overview of this fascinating topic, we present in Chapter 2 a description of essential elements of the scientific method. We make the case that, even though science is based on logic, progress in science is driven by insight, intuition, and inspiration. We also argue that the scientific community is diverse in its methods and approaches, and that this diversity is essential for making such progress.

Making the right choice of research group, adviser, and project is of crucial importance, yet students often flounder when starting a research project that is not the right fit for their interests and background, or when working with an adviser who does not meet their needs or expectations. Graduate students need to understand the considerations that should go into the choices of a research group, adviser, or research project. Guidance for making this choice, and other underlying ones, such as the choice of university, department, and adviser, is given in Chapter 3.

Because the choice of adviser can have such large bearing on the extent to which you do or don't *thrive* in your graduate study, in Chapter 4 we discuss the many different styles of advising and how these styles might or might not match with your personality and strengths (and weaknesses) as a novice in research.

We often think of research mainly as a process of solving problems. Most frequently, however, the largest scientific breakthroughs come primarily from asking the right questions, which comes about only after freely posing *lots* of questions, many of which will be off the mark. The role of asking questions is emphasized in Chapter 5. Once having posed questions that help frame your planned research, you measurably sharpen focus on the research path you will be following by organizing your effort into a formal (but alterable) workplan, as discussed in Section 5.3. The workplan is of critical importance when you are working alone, but even more so when your research is to be conducted within a group.

Chapter 6 offers advice on a number of systematic steps you can take to aid in the efficiency with which you carry out the workplan. Part and parcel of that workplan and how you go about carrying it out is the direction that you give to your work, founded on goals that you explicitly set, steps that you take in working toward those goals, and resources that you call on toward these ends. That chapter also reveals issues that bear consideration as you form your goals.

As in all aspects of life, the pursuit of scientific research is fraught with potential pitfalls. Also, as in life, when an apparent pitfall is recognized, the need to address it can offer an opportunity for a wonderful breakthrough. As amplified in Chapter 7, the key is in first recognizing the existence of the paradox or other sticking point and then realizing that it can be caused by a misunderstanding that you have had about the problem. The chapter continues with advice on making the most of confusions that inevitably arise during the course of research.

We might think that the purview of ethics and ethics violations is restricted to the professions of law, medicine, engineering, business, and politics. Not so. Each profession has its own specific codes of right and wrong behavior, in addition to those common to all fields of endeavor. Nothing can be more thoroughly damaging to a career in science than the loss of respect for the scientist's integrity and sense of fair-mindedness. The critically important topic of ethical standards in science and adherence to those standards is covered in Chapter 8.

Whatever the purpose of your research, it is essential that you dedicatedly follow the literature, not only at the outset, but throughout the course of all stages in your research. In Chapter 9 you will find efficient ways to use libraries and other modern sources of information.

The pursuit of science entails much more than defining a research problem, posing the right questions, and thereafter successfully solving the problem. Satisfying as having reached a solution might be, the scientific endeavor is incomplete until the outcome of the research is communicated broadly and well to the pertinent scientific community, and sometimes beyond to the lay public. Communication with clarity and in a manner that draws and maintains the interest and attention of the intended audience is an art whose development, for most people, requires dedicated care and attention. The task in communicating research results can often take as much time and effort as the research itself, as is emphasized in Chapter 10.

Knowing how to choose a journal that is most appropriate for your paper entails a number of considerations, and having an awareness of the submission and review processes (which vary from one journal to another) can help in minimizing consternation. Chapter 11 gives advice on these issues as well as tips for distributing your research results to the relevant group of readers.

Readily one can follow all of the guidance in the chapters reviewed above and yet fall into any of many lurking traps that divert time and attention needed to complete the research as efficiently as possible. Time management is yet another art form whose mastery eludes many people. Chapter 12 warns of time traps and offers ways of circumventing them.

Unless you are an independently wealthy renaissance scientist, you will need to secure funding for your research, if not while a graduate student then certainly in your subsequent career. Therefore, unless you have a sympathetic benefactor, the writing of proposals is in your future. Chapter 13 treats many facets involved in writing proposals, such as targeting an appropriate funding agency, knowing how to locate opportunities for funded research, knowing what are the elements in an effective proposal, and knowing what to expect in the proposal-review process.

A research project, whether conducted while in graduate school or in your career thereafter, ought not be considered an isolated entity. Rather, it can often purposefully be part of a larger goal or set of goals that define your research career. You can expect that your scientific career will follow a serendipitous trajectory that is recognizable only after the fact. You nevertheless can provide for a more satisfying and productive career by defining and working toward goals that you explicitly believe to be especially right for you, including those that are consistent with larger *meaning* that you wish to give to your life. Chapter 14 gives tips that can be useful toward that end, but expect that the best laid plans . . .

What will your career be? Depending on your personality, interests, and perhaps sense of mission, it could be one in industry or academia. Also in Chapter 14, we contrast careers, exposing both differences and similarities one might expect to have in industry, academia, and national laboratories. Moreover, frequently these days individuals choose to, or at least expect to, change their employer, their type of work, or function within the organization, and even their core career direction, for example, to a totally different scientific area (such as from biophysics to computer science) or from science and technology into management. Chapter 14 alerts you to such wide-ranging career issues.

Once you've successfully defended your thesis, you've almost launched your career in science. The next step, a key one, is successfully obtaining the job that you desire at the institution (university, national research laboratory, or private corporation) of your choice. Covered in Chapter 15 are a number of elements that enter into successfully obtaining that desired job. These include establishing contacts within the organization, the writing of a *curriculum vitae* (CV), how you approach the interview process, and factors to consider in assessing the organization and its practices.

Numerous books serve purposes that parallel those in this book. Some are listed in Appendix A. A sample curriculum for a course based on the material of this book is shown in Appendix B. Appendix C gives an overview of two common database formats for bibliographic references that are particularly convenient when writing multiple papers with overlap in their lists of references.

ACKNOWLEDGMENTS

A number of scientists visited the course SYGN501 at the Colorado School of Mines (CSM) from which this manuscript has grown. We especially thank CSM colleagues Dave Hale, Eileen Poeter, and David Wu for their creative and refreshing input. We thank Huub Douma for numerous interesting discussions, and for his constructive advice. Thanks to Lisa Dunn for educating our students, and us, on how to make optimal use of the scientific literature. Also, John Halbert and Pat Brie (both of Linguatec) offered helpful suggestions for the chapter on communication, pointing us to websites with tips to aid speakers in relaxing prior to and during their presentations. Julia Snieder made us aware of the wonderful story about the discovery of the periodic table of elements by Mendeleev. We are grateful for the many colleagues who have read earlier versions of this book, and especially want to thank the constructive criticism and suggestions from Chris Kohn (ExxonMobil), Susan Sloan (National Academy of Engineering), Nick Woodward (US Department of Energy), and Matt Lloyd (Cambridge University Press). The feedback from students who have taken the course "The Art of Science" has been important in shaping our ideas and in finding the right words. We are grateful to all of our students, and want to mention in particular Jason Deardorff, William Good, Matt Haney, Myoung Jae Kwon, Alison Malcolm, Ryan North, Russell Roundtree, and Yaping Zhu. We very much appreciate the critical and thoughtful suggestions by anonymous Cambridge University Press reviewers of two earlier versions of this book.

2 What is science?

Scientists, therefore, are dealing with doubt and uncertainty. All scientific knowledge is uncertain. This experience with doubt and uncertainty is important. I believe that it is of great value, and one that extends beyond the sciences. I believe that to solve any problem that has never been solved before, you have to leave the door to the unknown ajar. You have to permit the possibility that you do not have it exactly right. Otherwise, if you have made up your mind already, you might not solve it.

Feynman, 1998

Certainly a book for prospective scientists ought to explain what science is. Still, despite numerous books that treat the philosophy, character, and practice of science, there is no agreed-upon nor clear-cut and unambiguous definition of science. This holds not only for the many fields of study that have adopted methods patterned on those of the natural science, e.g., social science, psychology, economics, managerial science, and military science, but also for the natural sciences themselves. In a broad sense one might define science[1] as the activities aimed at understanding the world around us, but it could be well argued that the arts, humanities, and many other endeavors in modern society likewise aim at understanding of the world, albeit understanding of a different sort than that sought in the natural sciences. So let's focus on the practice of *natural* science, which might be defined as the activities aimed at understanding of the *natural* world.

The key word in this still broad definition is "understanding." This word, which itself is vague, has a number of differing aspects

---

[1] Science is understood by many to include the body of findings and understanding discovered through the practice of science. Throughout this book, where we use the word *science* we mean the *practice of science* as opposed to its findings.

FIG. 2.1. The painting "this is not a pipe" of the Belgian painter René Magritte.

that are found at the core of the scientific method.[2] These include the following:

- *A logical framework.* Adherence to the precepts of logic is fundamental to any activity that claims to be following the scientific method. For example, from the statement that "hoofed animals cannot fly" and the observation that "pigs have hooves," one can logically conclude that "pigs cannot fly." A conclusion that "pigs can fly" would either mean that the premise "hoofed animals cannot fly" is false, or that the observation that pigs have hooves is in error, or that one uses a system of thought that defies rules of logic. If the latter, then this system of thought would be inconsistent with the scientific method. The scientific method would, however, have a place in ascertaining the validity of the premise.

  In some endeavors other than science, there is nothing wrong with deviating from rules of logic, and, as shown in Fig. 2.1, those rules are not difficult to defy. The picture of the pipe (not a pipe?) satisfies a legitimate goal of a work of art in challenging and stimulating the viewer's powers of questioning – indeed of logic – and perceptions; art can spark inventive thinking on the part of the viewer. For example, it

---

[2] Much has been written about the intricacies of the scientific method (see Appendix A). Our discussion here will be kept at a broad level so as to emphasize a few key points as they relate to the meaning of *understanding* in the *practice* of science.

can reasonably be concluded that Fig. 2.1 truly doesn't show a pipe, but rather a picture of a painting of one.

Despite such quite acceptable departures from rules of logic, however, reasoning that is not firmly rooted in a logical style of thinking does not form part of the scientific method. This is not to say, nevertheless, that the scientific method always follows methodically or linearly from accepted paradigms in the sciences. The dramatic breakthroughs that initiate significant and exciting advances – sometimes 90-degree changes – in science have often been the result of free-ranging intuitive thinking by the most creative of scientists; we often call them *geniuses*. Such intuitive leaps, nevertheless, must always be followed up by thorough engagement along paths of logic and consistency.

- *A foundation in observations.* Science is no mere mind exercise of following paths solely of logical thought toward predictable consequences, mentally stimulating as that process might be. Mathematics is just such an exercise – typically a most demanding one – with the goal of arriving at indisputable truths, always however on the condition that its starting premises themselves are valid. Because science is aimed at understanding of the natural world around us, it needs something more than logic. As a result it arrives at vastly more than does mathematics or logic alone. At the same time, however, science cannot provide unassailable *proof* that its understanding and interpretation of the natural world is correct.

  Science's connection with the natural world is made through observations (today largely through use of sophisticated and sensitive instrumentation that vastly extends the range of our human senses) – typically painstaking and repeated, and often lifelong – chasing down implications and always seeking consistency across the observable world. Such observations can involve quantitative measurements, for example, measuring the normal body temperature of the highland gorilla. Others might be aimed at a simple yes/no answer, such as whether pigs have ever been observed to fly. (Note that the fact that no pig has ever been seen flying does not in itself *prove* that pigs can't fly,[3] but the contrary observation would indeed prove that they can.) The first type of observation is of a quantitative nature, the second a binary one. An observation need be neither quantitative nor binary; it can be more vague. For example, one might wonder to what extent a certain disease depends on nutritional habits. While such a dependency might ultimately be quantified by computing correlations based on observations, in early parts of the investigation more intuitive guidance would be necessary in order to establish this connection. Moreover, interpretation of the correlations themselves might be subject to the qualifications of uncertainty, verification of cause-and-effect, and clearing out of the way of various (sometimes many) competing factors.

- *Predictive power.* One of the great strengths of science is that from a known theory – possibly tested and calibrated with observations – one can predict physical[4] behavior in new situations. It is its predictive power that makes science of great use because it provides means for understanding, indeed influencing and changing (for the better, one can hope) the world around us.

- *Falsifiability.* All possibilities of error in understanding (or in a hypothesis) must be available for detection and evaluation; any hypothesis being tested "must be one that contains the seed of its own destruction" (Stenger, 2007). In the scientific method, the author of a hypothesis should do her utmost to try to "break" that hypothesis.

- *Repeatability and testability.* The methods and results of any scientific investigation must be amenable to being repeated and corroborated through independent testing by other investigators.

---

[3] Using logic alone, one could conclude that pigs might well be able to fly. They are hoofed animals, so if the starting premise were that "hoofed animals *can* fly," the *observation* that pigs are hoofed animals would lead one to conclude that pigs possibly could fly. The body of understanding in physics, however, indicates that pigs lack the equipment, e.g., wings or mechanical engines, and have an overabundance of mass-to-volume for them to be able to fly. Application of the scientific method would therefore lead one to conclude that it is highly unlikely that pigs can fly.

[4] Throughout the book, we shall use the word *physical science* as shorthand for all of the natural sciences, e.g., physical, chemical, biological, astronomical, geological.

To the above list of characteristics, we can add that science is challenging and difficult, features that are by no means unique to science. What makes science hard is worth elaborating and emphasizing, as we will do below.

Note that, contrary to the situation for mathematics and logic, in the description of science given here the word *truth* does not appear. The notion of truth implies that it is possible to establish once and for all (that is, to prove) that a finding or interpretation is *true*. This is, of course, a circular statement, but that is exactly the problem. In science, we have no *truth-meter* that allows us to establish that a given theory always holds with perfection. What we can do is compare the predictions of a theory with observations of physical phenomena. A favorable outcome does not, however, mean that the theory holds or can be validly used in every imagined situation. One can carry out a thousand measurements that support a theory, with no hard guarantee that the theory also holds for the 1001st measurement, especially when that measurement is carried out under somewhat different conditions.

The notion of truth implies the binary concept that something either is or is not strictly valid. In science one sees truth on a sliding scale. A case in point is mechanics. Classical mechanics, as originally formulated by Isaac Newton, has been and remains a foundationally powerful tool for addressing a broad range of phenomena in science and engineering. Yet it is not always accurate. After more than 200 years of success in predicting physical phenomena, it was found lacking (by Einstein). For bodies that move with a velocity on the order of the speed of light, classical mechanics needed to be replaced by the theory of special relativity. Moreover, small bodies at the molecular or atomic scale follow rules, seemingly bizarre at times, of quantum mechanics instead of classical mechanics. So is classical mechanics true? Well, this depends on the velocity and the size of body in which one is interested, as well as on the accuracy that one desires for predictions made with the theory. In science, it is more appropriate to speak of the *accuracy* than of the *truth* of a theory. Thus, the outcome of science can never be truth in an absolute sense; rather, science has the agenda of an unbiased search to discover ever more accurate understanding of the natural world. In the words of Moore (1993)

> *Erasistratus's beliefs, like all statements of science, are approximations to the elusive goal of "truth." Science is an accretive and self-correcting discipline and, generation after generation, its concepts become more precise and accurate.*[5]*

As mentioned above, much has been written about the scientific method and about the philosophy of science (e.g., Popper, 1965; Kuhn, 1962). Rather than attempting to give an overview of this topic, we present in the following two sections examples of different ways to categorize the scientific method.

2.1 DEDUCTION VERSUS INDUCTION

Induction is not an automatic procedure for advancing science. It depends on the brilliance, perseverance, knowledge, and luck of the scientist. And deduction is an effective and powerful procedure when one uses it to make testable deductions from provisional hypotheses.

Moore, 1993*

Scientific methods can be divided broadly into those of *deduction* and *induction*. The deductive approach follows the path of logic most closely. The reasoning starts with a theory that leads to a new hypothesis. This hypothesis is put to the test by confronting it with observations that either lead to a confirmation or a rejection of the hypothesis, again not the truth of the hypothesis but its accuracy under the conditions associated with the particular phenomena being studied. Portrayed in a flowchart, the deductive method thus takes the form:

Theory ⇒ Hypothesis ⇒ Observations ⇒ Confirmation/Rejection

Note again that agreement of the observation with the hypothesis does not mean that the hypothesis is always valid. For this reason,

---

[5] Erasistratus Chios was a Greek anatomist and royal physician (304–250BC).
* Reprinted with permission of Harvard University Press.

the confrontation of the hypothesis with observations can, in the most optimistic scenario, give support (sometimes, strong) for, but not proof of, the validity of the hypothesis.

An example of a deductive approach is the discovery of the neutrino, an elementary particle without charge and rest-mass. According to theory, mass, energy, and electric charge are conserved. It turned out that, with the decay products of beta decay that were known in 1930, mass and energy appeared not to be conserved during beta decay. This led Enrico Fermi to propose the existence of an electrically neutral particle without rest-mass. Because of its resemblance to the neutron, he called it the *neutrino*. In 1930, the presence of the neutrino was a hypothesis, and the hunt for its detection was on from that moment. It took until 1956 before the neutrino was detected (Cowan *et al.*, 1956) and its existence thus confirmed.

This example shows the amazing power of deduction. Logic applied to an existing theory led to a new and bold hypothesis – the existence of a new elementary particle. Finally the observation of the neutrino confirmed the bold conjecture of Fermi. It took an intense effort of 26 years finally to detect this particle, exemplifying that devising and carrying out definitive observations can be exceedingly difficult; one should not reject a hypothesis because the observations are as yet inconclusive.

In the deductive approach one works from theory to observations. In induction this order is reversed. Observations might reveal patterns that lead to the formulation of a hypothesis for an underlying cause for these patterns. The hypothesis, or a combination of hypotheses, can lead to new theory. The inductive method thus corresponds to the following flowchart.

Observations ⇒ Pattern ⇒ Hypothesis ⇒ Theory

An example of induction is the discovery, by John Snow in 1854, of the spread of cholera by contaminated drinking water (Goldstein and Goldstein, 1984). At that time, the existence of bacteria and viruses was unknown; how and why contagious diseases spread thus were not understood. Snow noticed a pattern in patients who contracted cholera; many of them had been drinking water from a particular pump in London. This led him to hypothesize that cholera had been spreading through drinking water, a major step that ultimately led Louis Pasteur to formulate the germ theory in 1857. The presence of bacteria and viruses was later confirmed by observations, thus establishing their connection with various diseases.

It might appear naïve for practitioners of medicine not to have realized that the spread of disease is caused by the presence of microbes in drinking water. This appears so, however, only in hindsight now that the existence of bacteria, and their role in spreading disease, has been firmly established. Without this knowledge, it was not at all obvious that disease should be spread by water: why would water make someone ill? It took an immense intellectual step to formulate the hypothesis that cholera was spreading through micro-organisms in drinking water.

These examples illustrate the power of both deduction and induction. Neither approach is better or more valid than the other. For any particular problem, one or the other of these contrasting approaches might be the more effective. In general, they are complementary; it is the combination and interplay between these methods that makes them so powerful. It could be said that many of the largest paradigm shifts in science have resulted from inductive, often intuitive, reasoning driven by observations. Examples include the quantization of energy (Planck) and natural selection as an explanation for evolution of species (Darwin).

2.2 REDUCTIONISM AND WHOLISM

Another way to characterize the different approaches is to divide scientific methods into those of reductionism and wholism. In reductionism one reduces a complicated problem into its constituents and aims to understand that complex problem through study of different components of the problem.

An example of a reductionist approach is the synthesis of protein. Protein consists of amino acids that are assembled in organelles in the cells called *ribosomes*. The order in which the amino acids are assembled is encoded by ribonucleic acid (RNA) after it has been copied from deoxyribonucleic acid (DNA), the genetic material in cells. The structure of proteins is incredibly complicated, but the reasoning above reduces it to an assemblage of amino acids. The assemblage itself is reduced to the coding of RNA, which in its turn is reduced to the structure and properties of DNA, which forms the blueprint of cells. Here, the large and complex problem, the synthesis of complicated proteins, has been reduced to simpler building blocks. And this reasoning goes further. DNA is a polymer that consists of four units, called *nucleotides*. The properties of the nucleotides can, in their turn, be described in terms of their chemical structure, further relating them to properties of their constituent atoms.

It might appear that the reductionist approach is the most useful because it reduces large problems to a combination of smaller ones that are easier to solve. This approach is, however, not suitable for every problem. As an example, consider an ant heap, an extremely complicated physical, chemical, and biological structure that is built and sustained by the collective effort of millions of ants. One cannot understand an ant heap by reducing ants to their constituent components, for example by dissecting them or by making NMR scans of their tiny brains. It is the complicated interactions among ants, as driven by their genetic material, the chemical and behavioral signals that are exchanged, and the way in which different types of ants are raised, that govern the behavior of the ants, leading ultimately to the construction of the ant heap. This is a problem that does not lend itself to a reductionist approach. The essence of the ant heap is its complexity, and one needs to view the ant heap as a whole in order to begin to understand its structure. Such an approach is one of *wholism*.

Just as with the distinction between deduction and induction, neither the reductionist nor wholistic approach can be considered universally superior or inferior to the other. It turns out, for example, that the reductionist approach for the synthesis of protein is not the full story: the interplay among different genes is essential for the activation or deactivation of specific genes. One cannot completely reduce the action of DNA to the action of isolated genes that spontaneously are copied into RNA. Moreover, genetic material is also passed on through mitochondria, the power plants in cells, that are transferred from mother to child. This is an example of a problem that seemingly can be handled well by a reductionist approach, but gains from wholistic elements for a larger understanding. A key reason that neither the reductionist nor the wholistic approach ought to be taken in isolation is that any view of the natural world (considered by some as *reality*) that a scientist devises is just a *model* loaded with assumptions and approximations of that world.

2.3 WHY SCIENCE IS HARD ... AND WHAT MAKES IT AN ART

When you take a step where there is no framework, you trust in your humanness. It's your humanness that gives you your edge. There are computer programs now that can do mathematics better than we can, that can arrange equations, and that can solve equations. Our edge is first of all understanding what it means, interpreting it, and going where that manipulation can't go because there is no logic step yet.

Weglein, 2003

As mentioned earlier, the scientific method is rooted in logic, but this does not mean that the path it follows is simply linear nor that the underlying logic is always easy to detect. In order to see this, let us view science as a collection of "facts." We introduce this rather vague term here to denote either a theory, a hypothesis, a pattern, or an observation. Thus, here, a *fact* is some element in the deductive or inductive method of Section 2.1.

Consider a scientific study that works through facts arranged in a linear sequence, as illustrated in Fig. 2.2. Here, the practice of science might appear to be simple – an application of the rules of logic

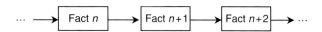

FIG. 2.2. Science as a logical chain of facts.

stringing together the facts. The situation, however, seldom will be so simple. While fact $n + 1$ can be a pattern inferred from observations that constitute fact n, it can be extremely difficult to recognize this pattern. Recall that it was not until 1854 that someone noticed that patients with cholera had been drinking from a well that was contaminated (the very notion of a well being contaminated was still unknown at that time). People had contracted cholera for thousands of years before that date, and the pattern with which cholera spreads had existed for a long time prior to its discovery in 1854. It took the imagination and creativity of John Snow to recognize this pattern. Similarly, in the deductive method, much creativity might be required in order to formulate a new hypothesis on the basis of a given theory. Sometimes, it takes courage to take the next step in science when devising a new theory or hypothesis that conflicts with "common wisdom" (e.g., how much "common sense" is there in quantum theory?). It can even take courage to carry out measurements that are controversial, such as when these observations could overthrow an accepted theory.

Science is hard for yet another reason. It can often happen that one of the facts – perhaps a theory or set of measurements – in the chain of Fig. 2.2 is wrong in that it conflicts with reality and yet the error is not obvious right away. It could be many steps down the chain of Fig. 2.2 before the error or inconsistency is recognized as being essential, requiring backtracking through the chain to find the erroneous link. The task of finding mistakes in earlier work can be extremely difficult, especially when the underlying assumptions being used have not been explicitly formulated. Beyond identifying a previous error, it also can take courage to acknowledge that one's earlier scientific finding might be incorrect.

The linear chain of Fig. 2.2 itself is a simplification. Often, it's not just a single fact that leads to the next, but, as illustrated in Fig. 2.3, a number of facts are required in order to take the next step. In this more realistic view of how scientific investigation proceeds, it can be difficult to determine that Fact E follows from a combination of several facts, say, A, B, C, and D. One might not know which of these

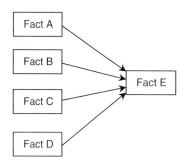

FIG. 2.3. Science as a complicated web of facts.

facts need to be taken into account in order to progress to the next step. Another complication is that, while Facts A, B, and C might be correct, Fact D could be in error. It often is no simple matter to discover such a problem and set the error right.

You might be surprised that we speak of facts that are wrong; this seems to be a contradiction. Keep in mind, however, that a fact that is assumed to be correct, but actually is wrong, is still a fact in the eye of the scientist pursuing the study. For yet another reason, errors are bound to be made. Consider once more the linear chain of Fig. 2.2. Where does the chain start? This is a deep metaphysical question that mankind has struggled with in various forms for thousands of years, and science does not tell us the answer. In science we aim to work our way down a chain, or web, of facts, but we don't know what was the first step. This is an unavoidable aspect of science that inevitably requires making assumptions. In investigations involving pure mathematics, a theory always starts with axioms, a set of "self-evident" rules that form the basis for the theory. In other fields of science, the initial assumptions are often not explicitly identified or even obvious. Always keep in mind that an assumption is nothing more than what it claims to be – an assumption. If it is invalid, conclusions based on that assumption are likely to be in error as well.

Considering the above discussion, while logic indeed forms the basis for stringing together facts, such as those illustrated in Figs. 2.2 and 2.3, strict application of logic does not always provide the rules for discovering the next step. Often such discovery requires imagination, creativity, intuition, inspiration, and sometimes courage. These qualities differ from just the ability to apply the rules of logic effectively. For this reason, science (always founded on or checked through observation) often is not a purely logical thought process. Because the qualities of creativity, intuition, inspiration, and courage, in addition to the ability to reason logically, are needed for doing original research, science is an art – an especially demanding and difficult one.

2.4 MANY WAYS TO PRACTICE SCIENCE

As we have seen, science is a mix of logic and intuitive approaches, so it should be no surprise that scientists have differing styles of working. Some are particularly strong at logic, but find it difficult to break out of this way of thinking in order to make bold leaps forward, while others operate in a much more intuitive way by relying on their gut-feeling, and afterwards use the logic necessary to justify the steps they have taken. Within the scientific community are many different approaches to research. Here, we sketch some of this variety, demonstrating that there is no uniform approach to doing science.

Some scientists choose to take a deductive approach and others an inductive one. The choice can depend on personal background, taste, and skills, but it can also be governed largely by characteristics of the specific field of research. In general, science with a strong theoretical emphasis, such as theoretical physics, tends toward the deductive, while scientists who deal with complex interconnected systems, such as ecology, gravitate to the more inductive. The distinction is not a rigid one. For example, in ecology, population dynamics often relies on mathematical models that follow a deductive line of reasoning. Most scientists use a mix of the two approaches; fortunately the scientific community is not divided between pure "deductors" and pure "inductors."

Similarly, some scientists heavily rely on a reductionist approach, while for others a wholistic approach is more suitable. Just as with the distinction between deduction and induction, the approach taken can be influenced by the taste of the scientist involved, as well as by the type of problem that is being investigated. In general, again, the reductionist approach typically works better for relatively well-defined problems that can be broken down into a limited number of simpler sub-problems, while the wholistic approach is better suited to problems involving complex systems with numerous interacting components.

Some scientists are drawn toward problems that require theoretical treatment, while others tend more towards challenges in experimentation. Even within these categories, one can work in different ways. A theoretical approach can involve mathematical work, numerical simulations, or literature studies carried forward by clear reasoning. Experimental work is not limited to laboratory experimentation; it could involve field work, patient trials, or other activities dictated by the purpose for which the science is conducted, e.g., curing patients, finding resources, or developing new technical devices. Note that both deductive and inductive approaches can involve either theory or observation, or a combination of the two. Ultimately, one needs to combine these different aspects of science: without a theory, observations are disconnected and their meaning cannot be understood, while theory without observations is a mind game ungrounded in our natural world. Many scientists gravitate solely toward either a theoretical or experimental approach, but often most progress is made at the interface of the two.

The scientific career can be pursued in different environments. Scientists do not work solely at universities, but also in industry and government laboratories. The type of institution in which the work is carried out tends to influence the character of that work. In general, research in academia is less focused on particular applications and is less restricted to producing specific deliverables; that is, it is more "academic." Research in industry is ultimately driven by the goal to make a profit. When the emphasis of such a research organization is on the short term, the science tends to be driven by the production of deliverables. Even within industrial research laboratories with relatively short-term targets, however, the research environment can offer the freedom, not to mention stimulation, of the best of university programs. Moreover, industrial laboratories with long-term goals have often given their researchers great freedom to push science forward without the need to develop applications that are profitable in the short term. One of many examples of such a laboratory is Bell Labs, which for a long time was a think tank noted for its innovative science.

A scientist usually wants to discover *how* and *why* things work. In contrast to this, the engineer's primary aim is typically to *make* things work. Thus, in general, it might be said that the engineer takes a more pragmatic approach than does the scientist. The distinction between science and engineering, however, is to a large degree artificial. In order to make things work, the creative engineer needs to know why they work. Similarly, in order to reach her goals, the scientist often has to construct devices, for example, by designing or building experimental equipment. Moreover, without a pragmatic outlook, the scientist in academia can readily get stuck unnecessarily. The distinction between science and engineering therefore is another example that operates on a sliding scale.

Thus, there is no single way in which all scientists (and engineers) work; one cannot speak of the generic scientist as one who approaches her task in a particular, narrowly defined way. As in other walks of life, diversity of style and approach, which is essential to the enrichment of the scientific endeavor and community, is based on the make-up of the individual. So, how you might be influenced by a particular scientist, such as your adviser, will not necessarily define the kind of scientist that you grow into and become. It is important to be exposed to the different ways in which scientists work, and to synthesize a personal style of scientific research that fits your skills, ambitions, and outlook. (This is sure to happen, in any event, whether explicitly or implicitly.)

2.5 WHY WOULD YOU WANT TO BE A SCIENTIST?

Just as there are many approaches to science, there also are many different motivations for being a scientist.[6] For many scientists, a natural

---

[6] There were no scientists prior to about the middle of the nineteenth century. But, how could this be – think of Newton, Hooke, Cuvier? Prior to then, those we know

curiosity is the prime motivation. This curiosity comes naturally for humans since it is its intellect that has made mankind so successful as a species. For many scientists, science simply is fun. The game of discovery and creativity is akin to solving a puzzle, with the potential reward not only of recognition by the community but of gaining further, deeper understanding.

Other factors motivate scientists. As mentioned earlier, science allows us to predict what happens if we do something novel, for example by building a new protein with novel properties in molecular biology, or by developing a new mathematical theory and using it for a new application. In general, society uses science to acquire power over the surrounding world. We use science to develop drugs that cure disease, to grow food more effectively, to explore and develop resources, to develop technology, and to find approaches that mitigate environmental damage caused by intensive application of technology. In short, science is useful, and for many scientists participating in the enhancement of this utility is a strong motivation.

For others, science means a career – an exciting and rewarding one. For those who are intellectually talented and highly disciplined, a career in science is a great way to make a living, whether at a university, in a government laboratory, or in industry.

Just as in our description of the various approaches to science, it is a mix of different factors that motivates scientists. However fascinating is the challenge of discovery, one still needs a job in order to pursue this challenge. For some scientists it is a mix of the pleasure and utility of science that is the main motivation.

2.6 WHO IS DOING SCIENCE?

The scientific community consists of a variety of different players. We focus first on those in an academic environment. Students often

of as pillars among scientists were individuals who followed what would have been called *natural philosophy* in pursuit of satisfying their elevated level of curiosity about how the world works, but they were not called *scientists*. The term was coined later.

divide the academic community into students and professors, but this is a gross over-simplification. Professors have differing positions and ranks, and, in general, play differing roles. Some professors spend most of their time and effort teaching, while other devote themselves almost completely to research. Many universities even have professors who do not teach at all. Some of these individuals are "on soft money," which means that they are not being paid by the university; rather, they generate their income solely through research grants.

Much of the research in some university programs is carried out by post-doctoral fellows, usually called *postdocs*. These researchers usually have just finished a Ph.D. degree and do full-time research that typically is funded by research grants. Since postdocs often have finished graduate school recently, their skills, knowledge, and recent experience can be a great resource for graduate students. Postdocs typically view their positions as essential for building the credentials necessary in order to attract offers for academic faculty positions.

Perhaps as much or more of the research at universities is carried out by graduate students who either work toward a masters (M.Sc.) degree or doctorate degree (Ph.D. or Doctor of Engineering). Research is an essential component of the graduate education. Graduate students learn by being trained on the job, led and aided by a faculty member, the *adviser*. As we discuss in Chapter 4, the adviser is the graduate student's principal mentor and coach, she or he plays a crucially important role in the success of that student's educational program and, no less, in the satisfaction that the student can have with the graduate experience.

Science is carried out at a variety of institutions other than universities, and, as shown in Fig. 2.4, the spectrum of activities in research and development is wide, ranging from fundamental research to the commercialization of knowledge. Fundamental research has no direct specific application; it is research that is driven by scientific curiosity "to know and understand" only. An example of this research is the detection of gravitational waves (e.g., Barish and Weiss, 1999). In contrast, applied research has the aim to extend existing knowledge

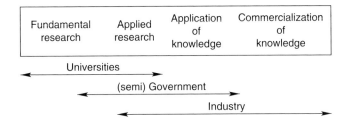

FIG. 2.4. The different stages of knowledge-based activities and the different players in the field. Modified after Speelman (1998).

to a specific goal such as developing a new product, or solving a societal problem. An example is the use of genetic engineering in the development of a new drug. The application of knowledge involves routine use of knowledge previously gained to a new and different set of problems. An example is the application of Geographic Information System to an existing database in order to map the urbanization in a region more accurately. The commercialization of knowledge has the goal to make a profit from existing techniques.

As seen in Fig. 2.4, a given type of research is not the sole province of one or another particular type of institution. The universities are predominantly concerned with both fundamental research and applied research, while the National Institute of Health, the Geological Survey and other (semi) government organizations have a focus on applied research and the application of knowledge. The activities of industry and other commercial organizations cover the spectrum from applied research (and even fundamental research) to the commercialization of knowledge. The boundaries in Fig. 2.4 are in practice somewhat fuzzy. For example, fundamental research on the quantum Hall effect was carried out in the Natlab, the physics laboratory of Philips. It is not clear (to us) whether that work should be called fundamental or applied research.

3   Choices, choices, choices

Any student contemplating graduate study, or embarking on graduate study or a career in science, is confronted by a myriad of choices. During your undergraduate career, you have the choices of major and minor subject, but you're likely past having made decisions on those. Once you have decided to pursue graduate study, near the end of your undergraduate study comes your choice of graduate university and program, founded on the choice of field – and, perhaps, subfield – of study you wish to pursue or type of career you wish to follow. Once beyond these choices, or perhaps concurrent with them, comes an all-important one, your choice of adviser – the individual who likely will have the largest influence on your approach to and outlook on science as well as on your success throughout the time of graduate study. An abundance of choices, many of which are puzzling and difficult to make, but what a wonderful position to be in to have created for yourself the opportunity to face such difficulties.

In planning research it is usually not hard to decide broadly what we want to achieve. The goal of the research might be to pursue an interest or a desire "to know"; it might be to find the path in which to establish a career; it might be to obtain a higher degree; it could be driven by the wish to contribute to making our world a better place in which to live; and it could be a mix of such considerations. None of them, however, tells us how to choose a field of research or program of graduate study, even less so a specific research project or thesis adviser. In this chapter we discuss considerations that could be of help in making choices that will have large bearing on a scientific career – the choices of university, department, program, and of research area and project. Because the choice of adviser is so all-important and has so many ramifications, we devote most of the next chapter to different styles of advising and how they might relate to your personality and needs.

Choices arise throughout a scientific career. Often, we might agonize and procrastinate in committing ourselves to making a decision, yet the ability and freedom to choose is a privilege that allows us to give shape to the future we wish to create. It might seem natural to choose the same field for your graduate studies as that for your undergraduate education. Although this most often is the path of least resistance, it is not at all necessary to continue in that same direction. It can be enriching to switch fields when entering graduate school. Why, at a young age, say your early twenties, ought you to be locked into a field for the rest of your life by choices previously made? Expanding into a new area of research broadens your horizons, exposes you to a different culture of research, and allows you to transfer skills learned in one field to another. This cross-fertilization of ideas and skills often is highly beneficial – for you, your fellow students and other colleagues, and for the new field in general. Switching fields, however, entails the task – with the opportunity – of making yet another choice.

Almost every field of research has specialized subfields, bringing on yet another choice that needs to be made, and, even when you know exactly in which subfield to continue, you are not done yet. Which university will you choose for graduate school? With which adviser will you work? And on which research topic will you apply your energy and talents? Decisions on these choices could be approached systematically and in sequence, but not always so. Perhaps you stumble into a research topic that excites you, or you encounter a particularly inspiring and charismatic researcher with whom you would like to work. Both sorts of often serendipitous encounters might well be defining for your future career.

We cannot offer a simple recipe for approaching the many choices nor for how to make them, and we certainly cannot tell the reader what decision to make. What we can do, however, is comment on considerations worth keep in mind while facing such choices. In Section 3.1, we present general ones applicable for a wide range of options in research. The many factors to consider when deciding on a university and department for graduate studies are discussed in Section 3.2, and, in Section 3.3, we suggest available resources that can be helpful in choosing an adviser. Section 3.4 on choosing a research project is divided into a number of subsections. Possible constraints that could influence your decision on a research topic are covered in Subsection 3.4.1. Additionally, as discussed in Subsection 3.4.2, research fields and subfields exhibit a typical growth pattern characterized by different stages over time. Given the current stage of any research field, in Subsection 3.4.3 we pose questions worth asking before embarking on a research project in that field. In Section 3.5, we offer thoughts on how well matched your personal skills and ambitions might be with a chosen path in research, and we conclude the chapter with a summary checklist of questions related to issues raised in the chapter in Section 3.6.

3.1 GENERAL CONSIDERATIONS

The effect of every action is measured by the depth of the sentiment from which it proceeds.

Emerson, 1841

Of primary importance is that your chosen research field offers the prospect of a career that you thoroughly enjoy. Our lives simply are too precious and short to aim otherwise. That is not to say that in your career you will encounter only tasks that you like doing; every job and every research project has its downside. Likewise, we don't advocate that your activities be driven primarily by selfish desires. Within the realm of research, however, the range of options is so wide that it is readily possible to find a research field and topic, as well as colleagues, to your liking; hence aim for a career that offers that probability. The pleasure that you derive depends on the content of the work, the people with whom you work, and the conditions under which you carry out your work. All these factors enter the equation when choosing your area of graduate research.

A major factor in the joy that you derive from your career is that your work be of interest to you. The best research is carried out by individuals who have a *passion* for their work. Sometimes passion is there at the outset; more often it develops over time and after some delving. The key to attaining that passion is that you know your field and subject in depth. You can expect to make a sound decision about the project of choice only after you get to know the research topic to some extent. This is a chicken-and-egg situation because time and effort are required to obtain background information on the research before embarking on a new project. Starting on a research project for which you have little affinity – perhaps because you were influenced by an adviser or colleague who was strong in steering you in that direction – can lead you toward a path of half-hearted effort, often with reduced promise of success and, ultimately, frustration. At times, you might find yourself in a situation that seems to offer little choice, but we urge that, despite the agony that you might feel, you seek to exercise *your* choice.

Although plenty of factors exist that could give rise to concerns, clearly our choices should not be driven by concerns or fears. You should be cognizant, however, of external conditions that can influence the satisfaction you derive from working in a chosen field and that sometimes can limit the opportunities to pursue a career in that field. There are economic and employment prospects to keep in mind, for example. How many jobs in your field can you expect to be available when you graduate? How much opportunity is there to move in a certain career direction by pursuing research in academia as opposed to industry, or to follow a career in teaching? Also, for a career in academia, what are the long-term and short-term chances of securing research funding? Some fields of research impose restrictions on the lifestyle of the researchers involved, for example by the need to attend to experiments during the night, or the requirement for extensive travel or fieldwork. These issues and questions often have no objective answers; those answers could well depend on your subjective expectations about future developments. Who knows what the job market will be five years from now, and how much funding will be available for research in a certain field after the next national election? While it is prudent to be aware of these external constraints on your research career, don't let yourself be driven by worries and concerns that might never materialize. You can pretty well be sure that if you have chosen a field in which you can find joy and love of what you're doing, your career will be one of success and satisfaction.

Whatever decisions you make, it is essential to be well informed beforehand. This usually requires some research, for example through searching the internet and talking with people, both well worth the effort. As a graduate or undergraduate student, it might be possible to do an internship in industry or participate in a summer research program at another university. These offer excellent opportunities to test the waters in an environment that differs from that in school and to become acquainted with a new research field, other research groups, and potential employers. Not only can such experiences help in making more well-informed decisions, the often serendipitous encounters during those visits can lead to new insights and opportunities that can be career defining.

3.2 CHOOSING A UNIVERSITY AND DEPARTMENT

On every level institutions can and should have a heart.

Pausch, 2008*

Assume for the moment that you have chosen a field for graduate study. Next, you need to decide *where* you want to do your studies. Perhaps you have already made and acted on that decision. In case not, again the first step is to become informed. Search the internet for the profile of different departments in your field of interest. Get the advice of people who are knowledgeable, for example by talking with your undergraduate adviser, professors whose classes you have taken, and graduate students who work in the same field. Considerations to keep in mind when choosing a graduate program can be summarized in the following list of questions.

- *What research opportunities exist?* What research is carried out in the department, and who is leading the research? What is the research productivity? What collaborations exist with groups at other universities and in industry? What facilities for research are available? Do students in the program receive funding support and encouragement to attend scientific conferences? What is the quality of the research compared with international standards?
- *How is the graduate program structured?* Not all programs are structured in the same way. Some require many courses, with little choice as to which courses to take, while others leave their students much more freedom to define their individual study program. How much research do students do? Is the research usually a team effort or do students tend more to work on individual projects? Are there opportunities to work or have exchange visits abroad?
- *What breadth of courses is available?* Courses taken at the graduate level offer a unique opportunity to deepen and broaden your knowledge and skills. What courses are being offered within the program? Be aware of courses offered in other departments; these can be particularly enriching and pertinent.
- *What is the quality of the faculty?* What is their research reputation? It usually is instructive to look at their publication lists because these are indicative of both their scientific productivity and emphasis. Pay attention to the list of authors of these publications. Are students involved, and, if so, are they listed as first authors? (As discussed further in Section 8.5, the first author is typically the person who contributed the most and received the most credit.) Are there indications of collaborations with others groups of high quality of which you are aware? Do the authors sometimes publish single-author papers? (This can be an indication of an active involvement in research.) Don't forget to seek advice from those who work in the same field.

- *What is required of the students?* The requirements of graduate programs vary widely. Many programs have comprehensive exams or qualifying exams, and the form of these examinations is variable. What are requirements for passing those exams? What is the time period within which they must taken, and how rigid is this time requirement? What sort of theses do students write? Must the thesis be written in the style of a self-contained monograph or can it consist of published or submitted research papers? If the latter, how close in theme must they be and to what extent must their messages be pulled together in the thesis? Must the graduate program be completed within a specific time period? Are there requirements to assist in teaching, and, if so, how much time and effort is required, and how much independence is given?

- *Are the program's requirements realistic?* The requirements imposed by a program might or might not be realistic. For example, students could be required to complete exams within two years of being admitted, but the program is structured in such a way (e.g., heavy course load) that it is virtually impossible to do so. It is worthwhile to learn of any such disconnect between theory and reality, and what have been ramifications for students in the program.

- *What are the costs and what financial support is available?* Tuition varies greatly among universities, and the cost of living is also highly variable. To avoid financial surprises, make an estimate of the expected costs over the duration of the various graduate programs. To what extent do research and teaching assistantships offered by the department or research group cover these costs? How secure is this funding and, for teaching assistantships, how much time are students expected to spend in teaching?

- *How long does it take to complete studies, and where do students go after graduation?* The time spent in graduate school varies greatly among different programs and often also between different advisers. How much time do students typically spend completing the degree? Be aware, again, of possible discrepancies between the official rules and reality. What fraction of Ph.D. students actually graduate with that degree? How quickly do these students find jobs, and where are they employed after graduation?

- *Are the students happy?* This is a simple question, but one of the best indicators of the sense of well-being of students in the program, and one that should weigh heavily in your considerations. When many students in the program convey a stressed or depressed impression, complain excessively,[1] or show other signs of discontent, the conditions for graduate study are likely to be less than desired. Our advice then is to search further and seek a research group where students indeed thrive.

Answering these questions will take thoughtful pursuit and persistent digging, but we cannot emphasize too strongly the importance to the quality of your graduate experience, and even subsequent career, of seeking and getting answers when choosing a graduate program. The time invested in researching these issues through seeking information and by talking with people will be well spent.

Students who are considering universities to which to apply and, if accepted, to attend, are understandably often attracted to the *big-name* universities. Those are typically the universities that rank near the top of various university rating systems. These institutions have earned these high ratings for a variety of reasons. Some of those reasons, however, might not relate closely to your goals for success in graduate study or might be inapplicable to your desired field of study. Even if a particular *elite* university previously had a strong program in your area of interest, for example, that might not be true at the time that you wish to start your graduate studies. This is worth your while investigating.

Most students apply to a number of graduate programs. Let us assume that you receive an offer from one or more of them. Before accepting or declining the offer, a valuable way to get insight into the merits and drawbacks of a graduate program is to visit the departments

---

[1] This is a relative issue. In conversations with a group of students some complaints are likely to arise, but in a healthy environment these are balanced by positive signals.

that have made you offers. This, of course, is not always possible, but when it can be done it provides the opportunity to talk with faculty and (especially) students, to visit the research facilities, to get a clear impression of living conditions and associated costs, and to develop a feel for the place. Also, as with most important decisions in life, it is best to let all information sink in for a while before making a commitment.

3.3 CHOOSING AN ADVISER

The next chapter covers, in depth, aspects of the role of thesis adviser and different styles of advising. Here, we focus on ways to get the information necessary to make an informed choice of research adviser and topic. While perhaps most students have not yet made their final decision on the choice of academic (i.e., thesis) adviser until after they have arrived in graduate school, for many students their choice of university and program was strongly based on the presence there of a noted professor in whose program they would dearly love to work. The suggestions in this section are equally applicable in either situation, although many are most comfortably and conveniently investigated once you've arrived at the university and are immersed in the chosen graduate program.

For a handy start, you can use the internet to learn about the research that individual professors are doing. The websites of individual faculty members can tell you in which areas they are carrying out research and might also convey their sense of enthusiasm and professionalism. It is also informative to look at their publication list. If you are seeking an adviser who is active in science, then you can expect to see an appropriate number of publications that reflect this activity. Be aware that abstracts for conferences are normally not seriously peer reviewed (see Section 11.2); such abstracts have a lesser standing than do publications in international peer-reviewed journals, but they nevertheless do convey an aspect of the professor's scientific activity.

Well beyond collecting this starting information, it is crucial to talk with many people, both faculty and students. Presuming that you are already in your new department, talk with its different various members, asking them about their research and projects. This not only gives you factual information, it also can indicate their level of enthusiasm (or lack thereof) and their communication skills. Is the person willing to take the time to carefully explain her research? Is she enthusiastic and motivating? It is a good idea to ask for material to read. This gives you the time to dig a bit deeper and digest the information at a comfortable pace. Likely, the content in the reading material will raise many questions. Returning with your questions to your potential adviser gives an opportunity to sense if she is willing to take the time to discuss your questions and provide understandable explanations. You need not have reservations about talking with faculty members. Most of them, particularly those with whom you might like to work, are enthusiastic to talk about their work and to engage with students. Academic faculty are paid to educate students, so sharing their knowledge and insights, even beyond the research, should be considered a part of the educational process. In the rare event that a faculty member is unwilling or unable to talk with you about research, this person is likely not to do so in the role of adviser either. It then behooves you to consider looking elsewhere for an adviser.

Sometimes, faculty give an inflated view of their accomplishments, as well as of the supervision and facilities they provide for their students. To get the inside story, it is best to talk with graduate students who have gained research experience with the various faculty members.[2] They can give you the student perspective on working in a particular group and with a certain adviser. Don't hesitate to talk with the more senior students; most of them will be delighted

---

[2] It can be an art to get honest and open answers in conversations such as this. It can help to ask open-ended questions rather than specific ones. For example, instead of asking "Is your adviser patient and understanding?" it can be more productive to ask "What do you think of your adviser?" When you wish to delve deeper into an answer, it usually is effective to start with an open-ended invitation, which offers the opportunity for the respondent to provide more detail. This can be done, for example, by repeating the last sentence that the other person has spoken, or simply by being silent. Since most people are uncomfortable with silence, they tend to fill the void, usually with additional information.

to share their experiences, including frustrations, with you. You ought not, however, take everything at face value; individuals might give an overly negative impression of an adviser because they carry a personal grudge, or they might be excessively positive because they are blinded by idolization.[3] It is therefore advisable not to rely on the opinion of solely one individual, but rather to integrate the information from several students into a coherent picture. It can be especially helpful to talk with the professor's former students. What were their experiences? How long did it take them to obtain their degree, and where are they working now? What percentage of the ex-students never finished a degree? It might not be easy to retrieve this information, but it can be an important component in your choice of adviser.

Be picky in choosing your adviser. When students are admitted to graduate school, they usually have been carefully evaluated by the department to which they applied. It is equally fair that the student also carefully evaluate the potential adviser. After all, the student and adviser will join forces for several years; both need to know as well as possible what they are in for.

3.4 CHOOSING A PROJECT
3.4.1 Will you have what you need?
A number of constraints should be kept in mind when choosing a research project. The best anticipated research topic in the most stimulating research group can turn into a disaster when you don't have the means to carry out the research. Therefore, you need to assess the resources that you require. Be sure that the facilities and infrastructure are adequate for the work that you intend to do. Your research might require expensive facilities such as a clean-room for experiments in genetic engineering or laboratory equipment such as a scanning electron microscope. Perhaps technicians are needed to build equipment for the research, or analysts are necessary to assist in large numbers of routine analyses or in measurements that require specialized skills

---

[3] Narcissistic advisers tend to attract such students.

or certifications. Perhaps your research requires large and unique facilities such as a neutron spallation source, or time on a marine research vessel that can be obtained only from national institutions or through interaction with other research groups, perhaps as part of an international collaboration. Some projects require large computational facilities for data analysis, numerical modeling, or data mining. The facilities available vary greatly among different research groups. Some groups are equipped with the most modern cutting-edge equipment, while others have substandard facilities that might limit the type and amount of research that the group can do, as well as the accuracy and reliability of the work produced. Substandard equipment can hamper work conditions for researchers, necessitating long repetitions for routine measurements that could have been automated, or night and weekend work because insufficient facilities are unavailable. In the worst of situations, inadequate facilities can even cause health risks for the researchers, although one would hope that such risks are adequately managed and mitigated.

You can be stymied if you pick a topic for which the resources are insufficient for carrying out the research, so it is prudent to have a clear picture of what you will need for your research. Be particularly wary of statements such as "data will be available," where the data have yet to be provided by an outside source. Delays in the availability of such data can cause unwanted delays in your research; also, too often such outside-source data turn out to be unsuitable for the proposed research. Likewise, be wary when you hear optimism that "a proposal will be funded" for purchasing equipment or to cover your tuition and stipend. With the intense competition for research funding, it is certain that a project is funded only once contracts have been signed.

Our intention here is not to scare junior researchers who are considering a career in science. Rather, we encourage you to be aware of your needs for carrying out a research project, and to be informed about the extent to which those needs are covered. At the start of a project, there might be no certainty on this issue; this is part of life, but it helps to be informed as well as possible before committing to a project.

Most research these days is carried out under certain time constraints. Funding for a project might be available for a limited amount of time only, a research sponsor might require results by a fixed date, and you don't want to spend more time in graduate school than is necessary. For these reasons it is crucial to be realistic in your time planning. Decide how much time you want to, or can, spend on the research and pick a topic that can reasonably be carried out within that time frame. It is difficult to plan the time required to complete original research accurately because innovative research is inherently unpredictable. Nevertheless, investing the effort into making a realistic time plan (see Chapter 12) for a project that fits within the limits imposed by the funding agencies, by your availability, and by the constraints of both the research group and your adviser can help minimize problems and frustrations later on. It can be helpful to define the timeline for intermediate milestones in the project, and to evaluate this timeline on a regular basis with the adviser.

3.4.2 The S-curve of development
Research fields usually go through a characteristic pattern of development over time. We illustrate this with the concept of the S-curve of development, which provides yet an additional ingredient in the choice of research area, topic, and project. Awareness of the different stages in the life cycle of a research field is beneficial because it has implications for the importance of the research and the character of the work out carried out in that field.

As a field of research develops a scientific idea or technological product, the returns on investments tend to follow the characteristic S-shaped curve shown in Fig. 3.1. You can find this curve in books of business management because it also applies, for example, to the development of a new aircraft or to the introduction of a new soft drink. In business, investment and returns are monetary concepts and are thus defined by their economic values. The concept

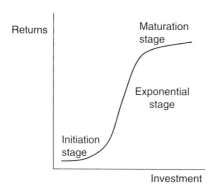

FIG. 3.1. The S-curve of development.

of the S-curve of development applies to research as well, but here "returns" and "investment" are defined differently from business. Of course, returns and investment do have an economic component in research, since research costs money and can lead to profit in the form of royalties or the profits of a start-up company that uses the results of the research. In research, however, the concepts of returns and investment are broader than in business. Investment in research includes the time required to carry out the research, the sometimes painful buildup of expertise needed to do the research, and possibly the opportunities lost by not pursuing some other line of research. In research, returns include papers or books published, the thrill of discovery, the fame and reputation associated with this discovery, the potential societal impact, and possibly economic benefits through royalties.

Let's discuss the S-curve of development in more detail. Whether we consider business or science, the principles are identical given the above definitions of investment and returns. At first, much investment is required and the returns are modest, with relatively little growth. Then once the product or research catches on, the growth in returns for the investment can be sizable. After a while, the product is well developed, and incremental returns diminish. In this model of the development of research, we can distinguish the following phases and what you, the researcher, can expect in each.

- *Initiation stage:* In this phase the research area is new. You don't have much experience yet, and the research methodology and infrastructure must still be developed. You are working on something original and might encounter skepticism from your colleagues and others in the field. For this reason it could be difficult to obtain research funding for your work.

- *Exponential stage:* Now the research has taken off. The tools and intellectual concepts for pursuing the work have been developed so that you can be highly effective in carrying out the work. The pace of scientific advance is high. Others have discovered this as well, so competition can be strong, but also many opportunities can develop for collaboration (especially when you are the one who had the good fortune to originate the initiation stage), which can add leverage to the pace and breadth of your current and future research. Research funding is easier to get than in the initiation stage. At some point, however, you might begin to question to what extent the research is still original since, by now, many research groups are likely pursuing similar work.

- *Maturation stage:* At this point everybody in the field knows the game. Much of what had previously been unknown, exciting, and new has by now been discovered. At this stage progress is incremental, and even to make this degree of progress requires the investment of large resources (notably the most valuable one: your time).

In choosing a research topic, it is helpful to have a good understanding of where your prospective project lies on the current S-curve of development. You might want to avoid embarking on research in the maturation stage because it offers neither much excitement, interest, nor glory. Some people might nevertheless choose it because it offers a sense of security or simply because they like the research in that field – an excellent reason for working in that area.

The choice as to whether you prefer your research project to be in the initiation stage or the exponential one depends on a number of factors.

- How good are you at generating new ideas?
- Do you want to be a leader or a developer?[4] Researchers with a strong urge to lead and tackle more risk-prone ventures often desire to work in the initiation stage, whereas those who are developers often feel more comfortable working in later, more accepted stages of a line of research.
- How dependent are you on attracting research funding for this project? Specifically, can you afford to start something completely new when you still need to convince research sponsors, many of whom these days have become growingly risk-averse, of the value of your work?
- Who are you? When you are an established authority in your field it is easier to initiate a new line of research than when you are a newcomer. If the latter, your collaboration with a respected adviser can be helpful toward establishing your reputation in the field.

Whatever choice you make for your line of research, if you work in a field long enough you will likely be involved at some time in research that is in the maturation stage. This does not mean, however, that your work in the long run is bound to become stale and that the return on the investment diminishes. It can happen that a research area that appears to be moribund becomes the springboard for initiating an exciting new line of research. This is indicated in Figure 3.2, which shows the S-curve of development for an old research line A heading toward a horizontal death of diminishing returns, but from which starts the S-curve for a promising new research line B that benefits from lessons learned from the old one. In this figure, line B, shown as a dashed curve, has associated with it a new set of investment and returns axes, also shown as dashed lines. Whether or not the fortunate inspiration for a new line of research springs from a mature one, for reasons of continuity of support it is desirable to initiate a new line of research before the old one reaches maturation.

---

[4] By *developer* we do not wish to imply one who lacks creativity, but rather one whose creativity is more directed toward filling in gaps and extending revealed concepts than blazing new trails.

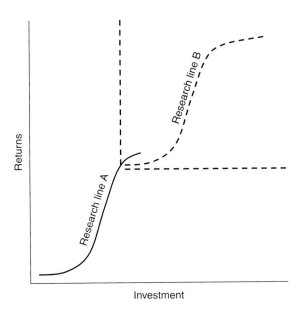

FIG. 3.2. The S-curve of development of two lines of research, line B springing from lessons learned in line A.

Examples abound of bright new ideas catapulting an apparently moribund mature science or technology into the initial stage of an exciting new advancement. An old line of research can grow into a new research direction in unexpected ways. One example is the field of paleobotany, the study of fossil plants. In the early 1990s, this line of research seemed to be focused solely on classifying and dating fossil remains of plants. Although the field was useful for dating layers in the Earth using the fossil plants embedded in the layers as markers, the methodology and applications of the research were fairly classical, and the field could be said to be *mature*. Plants absorb and emit gasses through small holes in the leaves called *stomata*. Later in the 1990s a correlation was shown between the density and size of the stomata and the carbon dioxide content of the atmosphere (e.g., Pigg and DeVore, 2001). Suddenly paleobotany played a key role in research on climate change because paleobotany could put constraints on the past CO_2 content of the atmosphere. From a field in the maturation stage, paleobotany suddenly transformed into one in the initiation stage by the addition of a new line of research – the reconstruction of paleoclimate. In this example, it was a new interdisciplinary application that gave the field its new-found importance. With the current emphasis on interdisciplinary research in science, many new opportunities for a rejuvenation and growth exist in new research.

3.4.3 Some probing questions before you start

Once you have identified a field that is still in development, that you find especially interesting, and that matches your goals for going to graduate school, you are not done yet. Unbeknownst to you, a problem that interests you might already have been solved. It will be highly frustrating to learn that, after much investment of your time, you have been addressing a solved problem. To minimize the possibility of this happening, do a careful and thorough literature and internet search, and talk with others to find out what is known about the problem, in particular if the problem has already been solved. In Chapter 9, we discuss efficient ways of conducting literature searches.

In practice, one usually does not just stumble into an interesting research problem without having carefully investigated the state of research in a field. That investigation is aided by posing and considering some pertinent questions. What is the big question that needs answering? What sub-problems are being investigated to help address this overlying research question? What is presently known about the topic? Are the accepted answers to the research question adequate and complete? (Many supposedly answered research questions were re-opened for investigation after the answers turned out to be wrong or incomplete, or when the advent of new data or measurement techniques provided new insights.) Are there any flaws in existing explanations? And, finally, once you understand these issues reasonably well, what do you see as open research questions?

If you have found a research question that interests you, and you are reasonably certain that the problem has not yet been solved, it is a good idea to ask yourself *why* the problem has not been solved. In particular, try to assess whether the problem is doable or whether it is so difficult that you cannot realistically expect to contribute much. Don't aim too low, but also be cautious not to aim excessively high. For any of a number of reasons, the problem might not yet have been solved. The basic research question could be flawed, for example, or the required experiments are considered not feasible. These are reasons for caution in embarking on a research plan. In contrast, it might be that a new data set is available that allows you to do research now that could not have been done in the past, that more powerful computers are now available, that new enabling analytical techniques have been developed, or that you simply have an idea nobody had thought of before. Sometimes the application of methods and technology in one field to another field opens up research possibilities that could not have been realized before. These are examples of reasons that could support your working on the unsolved problem you had in mind. An understanding of why the problem has not yet been solved therefore can be an important consideration in planning the research.

The sad truth is that we often do not reach the goals of a research project. This, however, does not necessarily mean that the project is a failure. You can make your research project more robust by asking yourself if intermediate results might be useful. Choosing a research topic that can produce useful spin-off results helps in minimizing the risk of embarking on the project. Moreover, being open to spotting both intermediate and serendipitous results can reveal interesting and promising new branches in the research. Sometimes it is possible to design an experiment in such a way that a negative result is meaningful.

3.5 HOW DO YOU FIT INTO THE PICTURE?

The right scale in work gives power to affection. When one works beyond the reach of one's love for the place one is working in and for the things and creatures one is working with and among, then destruction inevitably results.

Berry, 1992

Your research topic has the best opportunity for success when your skills and interests are well matched with the research. Posing to yourself the question "what is my edge?" suggests two choices. The first is to build on your strengths and branch out from there – the path of least resistance, at least in the short run. The second choice is to use graduate school as an opportunity to acquire skills in areas that are new to you. Here, you would be choosing to use graduate school to widen your horizons, and, by increasing your toolbox of skills, increase your prospects for finding a job after graduate school. In today's job market, it is increasingly valuable to have the ability and willingness to broaden your range of skills. The choice between these alternatives depends on your level of curiosity and security, on what you want to learn from graduate school, and on the amount of time that you have available for graduate school. In general, your research progress best thrives when you strike a balance between building on your strengths and learning a broadened range of skills.

Research in different fields can have highly varying character. Some fields are extremely competitive and developments are rapid, while the work style in other fields is more easy going. It is wise to choose a field that matches not only your skills, but also your personality, ambitions, and the amount of competition you are willing and able to face. Somewhat related, do you want to be a big fish in a small pond or a small fish in a large one? You should ask yourself what you believe you can contribute in a certain field of research and to what extent your skills and willingness to compete match the challenges and conditions in that field. A sort of pecking order exists in the competition for funding among different fields of research, governed typically by trends and fashions. (If you don't believe this, try your hand at estimating a pecking order for the following fields: genetic engineering, political science, geology.) In general, you will want to work in a field that, if possible, both satisfies your ambitions and affords the opportunity to excel. Be aware that the best short-term choice is not necessarily the optimal long-term one. It is easier to advance rapidly into a prominent career when you are a big fish in a small pond; if, however, the pond is too small for you, you might ultimately outgrow it and become frustrated. Therefore, keep in mind both the long term and the short term. Which alternative – small pond or big – works best for you depends also on your character, including your level of ambition.

Ideally, research leads within the allotted time to unambiguous results that provide breakthrough answers to pivotal research questions. In reality, this does not always happen. Measurements might turn out inconclusive; the problem might be too difficult to tackle; the measurements give clear but uninteresting or mundane answers; the work takes much more time than expected; or the project is more expensive than budgeted and needs to be shut down. Each of these possibilities poses a risk in research. Careful planning that involves a thorough literature search, ample brainstorming, perhaps some trial experiments, and the development of a research plan with a clear go-no-go timeline can help mitigate these risks, but a certain element of risk is unavoidably related to exploring new intellectual territory.

The risk level varies among different types of research project, and, as in business, this risk level often is correlated with the value of the (scientific) rewards. For some research topics, the outcome can be more or less predicted before the work is carried out, while for others the outcome is completely unpredictable at the outset of the investigation. When the outcome of research can be reasonably well predicted, the risk involved is minor but the chances of making a scientific breakthrough are commensurately slim. In contrast, lines of research whose outcome cannot be foreseen offer a distinct possibility that the work fails to lead to the results you had hoped for in the time allotted to it, but they offer the exciting prospect of scientific surprise and valued breakthrough. Different individuals feel most comfortable with different levels of risk and unpredictability. Through awareness of this issue, you can seek a research project with level of risk that matches your personality. As a graduate student you might want to discuss this issue with your adviser, fellow students, and others before committing yourself to a project.

A case can be made that innovative research *should* be highly risky. Consider the line of thinking and action advanced by Philippe Lacour-Gayet, Chief Scientist of Schlumberger, which pertains to research in an industrial environment, but whose essential point applies equally well to academic research: when a new line of research is initiated, the risk of success should be high for it to be worthy of the investment necessary to yield a ground-breaking new technology or product. After the concept has been proven and the feasibility studied, the outcome of the research is judged promising from a commercial point of view only once the estimated risk involved has decreased substantially. Then, only if the perceived risk is lower than a minimum threshold value should the product enter the development stage. The limitation on acceptable risk should thus drop throughout this sequence of stages, with product development proceeding only when the risk is lower than a minimum threshold value. If the expectations are being met, with the risk further reduced, production and commercialization can commence. By no means can one expect that the risk will drop monotonically or even at all; for example, a competing technology might come on the market that upsets the best laid plans. It is understandable that, in an industrial context, development and production should start only when the risk is below a certain minimum threshold, but in the concept phase, i.e., the research phase, the project is initiated only when the risk is *above* a perceived maximum threshold. According to the view of Lacour-Gayet, a new line of research should be started only when the risk is sufficiently *high*: it is worthwhile to invest resources in research only when it might lead to findings that are scientifically or technologically innovative and surprising.

You might or might not share this view on the merits of risk in research. Again, whatever your attitude about taking risks, it is worthwhile to be aware of the issues involved and that you seek an area of research that matches your willingness to take chances.

Realize that one can often design an experiment in such a way that the risk is mitigated. It helps to plan a project with care, and do a small feasibility study first. Designing an experiment so that intermediate or negative results can be useful and worthwhile helps in mitigating setbacks from stalled or inconclusive experiments. For risky projects it is especially useful to define intermediate milestones for the project, and times when the progress is to be evaluated in relation to these milestones.[5] One might even formulate a contingency plan for what to do when certain objectives are not met in time. Such a plan could include criteria for when, and how, to abandon the project. This helps avoid getting mired down in a project that fails to yield useful results.

3.6 A CHECKLIST WITH QUESTIONS

In this chapter, we have covered many different issues worth taking into account when choosing a research topic. Here is a summary list of questions that you might pose to yourself in making that choice.

- Do I have a genuine passion for the research project?
- Do I really like my research supervisor and my close colleagues?
- Do I have the resources that I need?
- Is the time planning realistic?
- Where is the project on the pertinent S-curve of development currently? Can it potentially initiate a new S-curve?
- What is my edge, and does the project match this edge?
- Has the problem already been solved?
- If not, why has it not been solved and why do I think that I can solve it now?

[5] This approach is more the rule than the exception in industrial research, where management carefully balances the investment in research with the outcome of the work and its usefulness for the organization.

- Can intermediate results be useful?
- Do I want to be a big fish in a small pond or a little fish in a large one?
- Are there adequate job opportunities and funding options for the chosen field?
- Is the project on the one hand sufficiently risky to be worthwhile, but on the other not so risky as to lack promise for success? What are the potential rewards given the risk?
- Do I have all the information needed to make an informed decision?
- What is the value of the project?

As a student you might consider discussing this list of questions with your adviser, and with other professors and fellow students, before embarking on a research project.

4   The adviser and thesis committee

Throughout their careers, scientists work with others. This is particularly so for students during their time in graduate school. As a graduate student, among those with whom you can expect to interact variously throughout your graduate-school career – your academic adviser, faculty members on your thesis committee, other faculty members from both within and outside your home department, fellow students, and scientists elsewhere whose work and ideas can be of value to you – it is your academic adviser (thesis adviser) who can be expected to influence most directly the course, for good or ill, of your graduate experience. In Chapter 3 we discussed the many choices that researchers, in particular, beginning graduate students, must make in order to focus their research and, for graduate students, to complete their educational and research program in a reasonable amount of time. The choice of the adviser and the related choice of a research topic are essential not only for the successful completion of graduate studies, they can also influence the way in which the academic career develops, and even the course of one's subsequent scientific career. In order to make informed choices, it is necessary to understand the roles – both academic and personal – of the scientific adviser, with whom you can expect to spend many hours. We describe those roles in this chapter, along with the related role of the thesis committee.

4.1 DIFFERENT STYLES OF ADVISING

The satisfaction and success of research are determined by more than the content of our research; they also are largely influenced by the people with whom we work. Suppose you identify a research project that you find attractive, but your direct colleagues for the project are people who provide little or no inspiration. Would you then look forward to going to work every day? Could you expect these people to encourage you when things don't work out? Would such a group of people provide you with the sounding board that you need for feedback on your ideas and on the challenges you face? Research in such an environment is unlikely to be as successful and as gratifying as that in an environment with people who inspire you and whom you truly respect and enjoy being with.

In graduate school this issue takes on particular importance. A graduate student depends strongly on his or her adviser for guidance on both academic and research decisions, but often on personal ones as well. The adviser is the person who eventually might be making the most important decisions necessary for completing a degree. Much more, ideally the adviser also provides the context and the means for the research, acts as a coach in your academic development, inspires and challenges you, and encourages you when you need it. It is therefore of crucial importance that you give thought and attention to choosing a supervisor (and co-workers) whom you like and respect. The thrill of uncovering a scientific truth or of watching your ideas fall into place can be incomparable. If, in addition, that thrill is shared with inspiring collaborators, such discoveries can lead to further ones at an exhilarating pace. Moreover, although it always entails hard work, digging, diligence, and concentration, at its best research can be joyous and fun. In contrast, when the chemistry between you and your adviser is not right or, for whatever reason, you cannot communicate well with your colleagues, you can expect collaboration to be impaired.

Not surprising, as with all other human beings, advisers are imperfect. Lelieveldt[1] (2001) encapsulates the characteristics of different types of adviser in the following list.

---

[1] Translated from Dutch by the authors with permission from Lelieveldt.

The unavailable is virtually never there, and if he is there he is "busy, busy, busy."

The magnum opus writer is not concerned with the limited time that you have to write your thesis: "I also took ten years to get my Ph.D."

The critic knows precisely what's wrong with your text, but makes no suggestions about how you could modify it.

The nitpicker corrects minor grammar errors (preferably with a red pen), but has nothing to say about the content or structure.

The structuralist hammers on about structure, but has nothing to say about the content.

The indifferent has no opinion, thinks everything is OK.

The hyperactive sends you repeatedly into a new direction with his wild ideas, caring little about coherency and direction.

The conservative holds on to what he has grown up with, abhors everything he is not familiar with.

The talker does most of the talking without listening to what you have to say.

The imposer has strong ideas about what your research should be like and accepts no deviations from his ideas.

The competitor feels threatened because you might soon surpass him; therefore he puts on the brakes.

The professional has ample time for you, reads your manuscripts promptly, gives adequate comments, motivates you, and combines the roles of coach and evaluator in an admirable way, and is thrilled to see your ideas forge ahead of hers.

The personalities sketched in this list are caricatures, although various of the characteristics are not far from the mark for too many advisers. Clearly, the perfect adviser is *the professional*. Alas, perfection is hard to come by in this world. Happily, few advisers completely fit one or more of the other categories. The list, nevertheless, offers useful pointers for identifying possible strengths and weaknesses of advisers and can help you spot character traits that indicate whether a professor would likely be effective or not for supervising your project.

What other character traits might you wish to find in an adviser? Here is a further list of traits to look for. The perfect adviser:

- is professional, with a good reputation in her field of research,
- is experienced and creative in research and in the supervision of graduate students,
- is enthusiastic and knows how to motivate others,
- has the time to help students with their research and stimulate them in other ways,
- challenges her graduate students and stimulates them in their growth,
- has enough research funding to provide financial support for students and the infrastructure needed to carry out the research effectively, and
- has good human skills. Some tend to believe that science is a purely rational and intellectual activity, but it helps greatly when you can share the joys and sorrows with others in the laboratory.

This is quite a list, probably causing you to wonder if the super-human actually exists who fits the bill. In reality, you would be remarkably fortunate to find an adviser who has all or most of these character traits. This list nevertheless can be useful as you seek the adviser who well fits your needs.

There is seldom such a professor as the "generic adviser", who has only one fixed way of working. The various ways in which professors work can suit the personalities of students differently. Best is when the style of advising matches the needs of the student. Of importance is the amount of time that advisers spend on their students, as this is highly variable among advisers. Some students hardly ever see their advisers, while others talk with them almost daily. One of our colleagues actually shares a large office with his students, to ensure there is no barrier to communication. Clearly, the adviser who hardly ever meets with her students is unlikely to be effective for offering the help and supervision the students need. This does not mean that the other end-member, the professor who meets

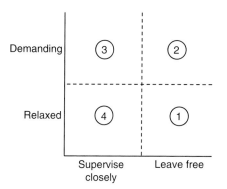

FIG. 4.1. Four quadrants with different styles of advisers.

frequently with students, is ideal because in such a situation those students might lack the freedom to acquire the scientific independence and problem-solving skills necessary for growth in graduate research. No hard guideline for the amount of time that a graduate student needs to spend with the adviser can be given because that need depends on the experience level and level of technical skills, as well as maturity, of the student. Whatever the advising style, however, enough time should be spent together for the student to learn from his professor.

Some advisers give their students wide latitude and freedom to pursue their own research ideas, while others supervise their students closely. This is indicated along the horizontal axis of Fig. 4.1. Neither of these approaches is best in all situations. Some students need close supervision because of their inexperience, while those with more scientific maturity are better off when left free so that they can more fully use their creative powers and grow toward independence. The degree of supervision that works best depends both on the student and her needs, and on the style of the adviser. One of our former colleagues left his students completely free – a "sink or swim" style of advising. He chose this approach not from lack of caring, but because he genuinely believed that students should acquire their scientific independence at the earliest stage. The downside of this approach was that about half of his students left within a year; those who survived were extremely successful in their scientific careers. This style would be too severe for many young students, but perhaps ideal for others. Although it won't be easy and obvious for you to know just which style of advising would be better for you, it certainly is in your interest to be aware of an extreme of this sort when considering under which professor's guidance you envision studying. The tough approach is certainly not our style of advising (in case you'd like to work with either of us), but the example shows that some professors can have quite extreme viewpoints on approaches to advising, any of which will have both advantages and disadvantages.

The vertical axis of Fig. 4.1 shows yet another aspect of different advising styles. Some advisers are demanding, while others are more relaxed and easygoing. It might seem that the latter type of adviser is preferable, but this is not necessarily so. Advisers can be demanding out of a genuine desire for the student to be successful. An adviser who goes out of her way to create research and career opportunities for her students, might expect a similar effort in return from these students. Again, neither style of advising is uniformly more effective than the other, but it is best when the advising style matches the character and needs of the student. A high-strung student might become even more tense when working with a demanding adviser, while a student with a relaxed attitude might better respond to an adviser who clearly states expectations and requires the student to deliver in a timely fashion.

The four quadrants in Fig. 4.1 indicate combinations of the different advising styles. The perfect adviser adapts her advising style depending on the needs of the student. For example, an inexperienced student with a casual work style might need to be supervised closely, with clearly stated expectations (quadrant 3), while an inexperienced, insecure, and tense student might be better led by a relaxed adviser who offers close supervision (quadrant 4). The highly talented and creative student with a tendency to be distracted probably thrives when supervised by a demanding adviser who otherwise leaves the student to give shape to her research (quadrant 2), whereas the equally talented but tense student could be best off with a relaxed adviser who allows considerable freedom (quadrant 1). The advising style might well change during the course of a Ph.D. project. As the student gains experience over time, the need for close supervision could be lessened. When approaching completion of the thesis, most students have a strong internal drive to finish and thus do well with a relaxed advising style, but a minority of students continue to need clear deadlines and close oversight in order to wrap up their project. Those students are likely to benefit from a demanding adviser.

While the ideal adviser switches from one advising style to another depending on the needs of the student, many don't have the skill, insight, and patience to do so. As a result they rely on just a limited skill-set of advising styles. This is not a problem when that advising style matches the needs of the student, but when it doesn't, research progress, growth, and well-being of the student can be jeopardized, to the detriment of both student and adviser. For the above reasons, we encourage beginning graduate students to be aware that choosing an adviser should involve more than just that of finding one who works in an area of mutual research interest. Compatibility of personality, needs, and outlook for mode of working can be of comparable importance. *Be aware of these issues and become well informed before choosing an adviser.*

Whatever the advising style, it is reasonable to expect and demand that you learn much from your adviser – that she serves as both teacher and mentor. As a graduate student you work hard for little, if any, financial gain. This is fine as long as you enjoy your work, and you learn and grow towards scientific independence. The teaching role includes not just the technical skills of doing research, but also the practical aspects of being a researcher. The mentoring role plays an essential role in helping the student build a well-balanced and successful career. When delivered with care, attention, and compassion, the advice and feedback passed on from the adviser helps define and promote the growth of graduate students into professional, independent researchers.

The appropriate qualities of supervision are considered to be so important by the American Physical Society (APS) that the APS Council adopted the following statement on treatment of subordinates:

> Subordinates should be treated with respect and with concern for their well-being. Supervisors have the responsibility to facilitate the research, educational, and professional development of subordinates, to provide a safe, supportive working environment and fair compensation, and to promote the timely advance of graduate students and young researchers to the next stage of career development. In addition, supervisors should ensure that subordinates know how to appeal decisions, without fear of retribution.
>
> Contributions of subordinates should be properly acknowledged in publications, presentations, and performance appraisals: In particular, subordinates who have made significant contributions to the concept, design, execution, or interpretation of a research study should be afforded the opportunity of authorship of resulting publications, consistent with APS Guidelines for Professional Conduct.
>
> Supervisors and/or other senior scientists should not be listed on papers of subordinates unless they have also contributed significantly to the concept, design, execution, or interpretation of the research study.
>
> Mentoring of students, postdoctoral researchers, and employees with respect to intellectual development, professional and ethical standards, and career guidance, is a core responsibility for supervisors. Periodic communication of constructive performance appraisals is essential.
>
> The guidelines apply equally for subordinates in permanent positions and for those in temporary or visiting positions."
>
> Kirby and Houle, 2004

This APS statement of dos and don'ts in the treatment of subordinates provides a comprehensive and useful list of legitimate expectations you can have of your adviser.

While you have much to learn from your adviser, in a proper relationship between adviser and students, the word "subordinate" should be taken in only a limited sense. The most satisfying, rewarding, and productive relationship is one in which the adviser has the qualities and character of a role model, becomes your mentor, and is a *friend* – perhaps for life. Also, the best of adviser–student relationships is one in which both are partners in the discovery process. Your adviser clearly has the edge in terms of breadth of experience, technique, and understanding of the context underlying a research topic. You, in contrast, have the edge of naïvete and perhaps of energy and enthusiasm. You can come up with the fresh "why not?" questions that can be the springboard for breakthroughs on problems previously presumed to be insoluble. Initially, your adviser is your teacher; perhaps as your research develops, you can become his or her teacher as well. Nothing excites the professional adviser more than when this happens. This excitement is captured beautifully in the following quote from Jacob Bronowski:

> *It is important that students bring a certain ragamuffin, barefoot irreverence to their studies; they are not here to worship what is known, but to question it.*

Despite the best of forethought and intentions in choosing an adviser and research project, it can happen that something about the relationship between you and your adviser is amiss. It could be an apparent dead end in the research, something about the research that turns out not to appeal to you, or poor chemistry in your relationship. You should not feel that you belong to your adviser; you are not an indentured servant. You should feel free to discuss your concerns with your adviser, with other faculty members, or with members of your thesis committee. With good reason for doing so, you ought to have the freedom to change advisers. Although you might understandably be concerned that investigating a change of advisers could hurt the feelings of your adviser, this should not be an overriding factor in your effort to set this right in your graduate program. In seeking a change of advisers, you of course must find another professor with whom you would like to work and who would like to have you as a student, and a research project that appeals to you and for which that professor has funding.

4.2 THE THESIS COMMITTEE

Depending on the country, many universities require that a graduate student have a thesis committee. The thesis committee, typically composed of professors from within the home department and from other departments and even outside the university, serves a formal purpose in that it must agree with and sign off on student proposals for courses taken in graduate school. It must evaluate the progress of the student; it often administers and judges performance in comprehensive or qualifying exams; and it evaluates the thesis and thesis defense. These are important formal tasks, and it is essential that a thesis committee carry out these tasks with attention and care. For many students, the contribution of the thesis committee is limited to these tasks. If it fulfills only these formal responsibilities – large as they might be – the thesis committee, we believe, would be one of the most under-utilized resources of graduate students. When used more extensively, the thesis committee offers a valuable source of help and advice for graduate students.

Rather than viewing the thesis committee solely as a group of examiners who need to provide signatures on essential forms at the right time, you would do well to view it as a body of experts who are willing to help broadly with your graduate studies. Members of your thesis committee can act as a sounding board for your ideas, conjectures, and plans. This can either be in a formal role, as when you meet with the full committee or, on a smaller scale, such as when you might meet with just one or two members of the committee. This makes it possible that you gain valuable advice, insights, and points of view from experts that go beyond those of your adviser; it can even lead to worthwhile scientific collaboration. It matters little whether these meetings are formal or informal, or whether you meet with the entire committee or with a subset of its members. Choose the format that works for you, and use committee members as a valuable resource for your research and career development.

Members of the thesis committee can also act as informal mentors, and their coaching and advice can augment that of your adviser in helping steer your development as an independent professional. Their network of colleagues can be useful for your research, for obtaining internships, and possibly for future career opportunities. Apart from their role as mentor, committee members can help in conflict resolution between student and adviser. We hope, of course, that such discord does not arise, but even when working with the best intentions, conflicts do happen occasionally. Just as it is best when you and your adviser consider yourselves as friends, the same holds if you can develop a similar relationship with committee members.

Since the thesis committee can play such a valuable role, it is important that its members are chosen with care. Most academic departments have regulations that govern the composition of the thesis committee, but these rules usually offer some freedom concerning who to choose for the committee. Members of the committee should have the professional expertise in your general research area necessary to understand your work and provide meaningful feedback. For this reason professors from your home department often serve on the thesis committee. Keep in mind, though, that faculty from other departments or researchers from other organizations, such as industry, can have valuable, sometimes even better-targeted, expertise. Committee members from outside the department can provide unexpected and refreshing insights on your work, and they often have a range of contacts that goes beyond that reached by committee members from your own department. This argues for seeking a diverse thesis committee. Also, for reasons given above, you hope to get along well with the members of the thesis committee. Therefore, so as to facilitate open and enthusiastic communication, choose members with whom you have a personal rapport.

Be aware that, at most universities, it is easy to make changes in the composition of the thesis committee. When assembling the thesis committee, you often don't initially know its members well, and with time you might discover that some members don't fit your personal or professional needs. Also, as you are carrying out the research, it might become apparent that it would be valuable to bring new expertise into the thesis committee. Don't be hesitant about making such changes when you feel that is helpful. The thesis committee exists to serve the student, so it is perfectly acceptable to make changes that improve its usefulness for you. When these changes are made for good reasons that can be explained tactfully, most committee members will accept changes gracefully.

In summary, we encourage you to see the thesis committee in a broader context than as a group of professors who merely provide the signatures needed for getting you through graduate school. See them as a rich source of expertise, use them as a sounding board, and view them as valuable mentors.

5 Questions drive research

In the theoretical physics community there are many more people who can answer well-posed questions than there are people who can pose the truly important questions. The latter type of physicist can invariably also do much of what the former can do, but the reverse is certainly not true.

Zee, 2003

Many scientists (and non-scientists as well) live under the impression that they don't know much about research topics that lie outside those that occupy them on a daily basis. We often feel like a blank sheet of paper when it concerns such research topics. Perhaps we know more about various areas than we think. The path to ferreting out aspects that we do and don't know is lined with questions. For example, both of us authors are geophysicists, and (an understatement) we don't know much about biomedical research. Yet if we take a topic such as cell therapy, several statements (correct or incorrect, naïve or otherwise) about this area of research can readily bring a number of questions to our lay minds. A line of thinking might go as follows.

> In cell therapy one seeks to modify the genetic material of cells in a body in order to correct the deviant behavior that causes a disease. The genetic material is stored in large molecules called DNA. Viruses modify the genetic material of cells. *Can we use existing viruses for cell therapy? How do we change the genetic material of viruses? Are there ways other than through use of viruses to change the genetic material of cells? We know of such an activity as genetic engineering. Must that technique be used in the laboratory only, or can it be applied in-situ?*

This list of what we already know and questions that they inspire can readily be made longer and progressively more sophisticated. This, of course, in no way implies that we have any degree of expertise in the field of cell therapy (quite the opposite), or that we can formulate research questions that are pertinent for front-line research in cell therapy. Even though many of our notions about the subject are likely erroneous, containing many misconceptions, we nevertheless are not blank sheets of paper regarding cell therapy or many other subjects.

The example above suggests that formulating questions is the precursor to further learning. Posing questions is such an essential part of innovation in research that we focus on the role of questions in this chapter. But where do the questions come from? When embarking on a new line of research, it is useful first to make a list of what you know, or think you know, about the research topic. You need only take a fresh piece of paper, or fresh computer page, and write down or type everything you know about the topic. Don't be critical and don't "filter" your ideas before you write them down. It's as useful to discover your misconceptions as it is to find out which of your notions are correct. This starter list of unfiltered ideas fertilizes the soil from which questions sprout.

5.1 THE NEED TO ASK QUESTIONS

He [Aristotle] provided a pattern of learning which is among the most difficult of all steps in science. It consists of being able to ask questions in such a manner that data can be sought for an answer.

Moore, 1993*

That is the essence of science: Ask an impertinent question, and you're on your way to a pertinent answer.

Bronowski, 1973

Suppose you are lost in a forest and you need to find your way out. One approach is just to go off in some direction and hope that this gets you out. If, however, you don't know where you are going and have a hard time keeping your bearing, this usually leads to endless wandering through the forest, getting nowhere. Instead, what might you do? You could ask yourself questions such as: *Can I see any*

---

* Reprinted with permission of Harvard University Press.

landmarks, for example, a hilltop that will allow me a view over a larger area? Is there a river or a road that could lead me to population? Is there any direction, such as downhill, in which I can reasonably expect to find a way out? Are there any signs of human activity such as traffic noise or smoke? From having posed these questions, you can start to devise a strategy for getting out of the forest.

Consider another problem. You are developing software and are stuck with a bug in your code. How do you find this bug? You could just stare at the code long and hard. This usually leads to myopia rather than to quick results. A far more efficient way to uncover the bug is to ask yourself questions aimed at helping you find it. (This might sound like circular reasoning, yet this is the way to proceed. Before you can find it, you must first focus on where to look and what to look for.) For example, can you narrow down the bug to specific parts of the program, such as a particular subroutine? In order to answer these questions, you can insert test statements in your code or write a test program. In the short run this costs time, but over the long run it can save a great deal of time; it often is the only way to find the bug (unless you are lucky).

The upshot of these examples is that, to solve your problem, the key is not to look for answers but to ask the right questions. If you pose enough questions, some are bound to be right ones. Asking questions is essential in research because they help give focus, and without such focus we are groping in the dark. Questions posed and later pondered give needed direction to our research.

What would our world-view be like, for example, if Einstein had not asked himself the question "what would happen if the speed of light is the same for every observer?" At first sight this seems an absurd question because it implies that two observers moving with respect to each other see light propagate with the same speed. Yet if Einstein had not thought to pose this question, the theory of relativity would not have been formulated, the equivalence of mass and energy would be a puzzle, and many technological advances, such as satellite navigation (Ashby, 2002) would have been impossible. Much of our world-view and technology have been shaped by great minds who were motivated to pose and follow up on questions that at first sight seemed outlandish, but whose inquisitive nature led to discoveries that could not have been imagined before.

Perhaps surprising, when you ask the right questions the answers usually follow rather directly. But, are all the questions you ask the *right* ones? Of course not, so pose lots of them. The process, however, works only when the questions are sufficiently specific. Suppose you are still lost in a forest. Asking yourself the question "how can I find my way out?" is not of much help unless it is used as an aid to thinking of more well-defined following ones. The questions must be sufficiently clear-cut that they lead to actions that will guide you to the problem's solution. If the questions fail to do this, you have to make them more detailed. This process of asking yourself a succession of questions that are progressively more specific could be considered as a path that might lead you out of the forest. Although subsequently finding the answers to these questions could well entail much work, such a path usually is the only practicable way to start toward solving your problem.

Being innovative in the type of questions that we ask is essential. As stated by Nick Woodward, program manager at the US Department of Energy: *"we need to change the nature of the questions we ask, not just seek better answers to the questions we already have."* Wouldn't it be good if all the questions you asked were the right ones in some sense? While that, of course, cannot be guaranteed, a prime characteristic of the right sort is that the question leads to tangible action. Its promise of accomplishing that, moreover, requires that the question not be so large that the suggested course of action requires too many steps in too many directions at once. Breaking up the large question into smaller ones usually helps in formulating those that lead to action that ultimately helps solve the problem.

It is often said that scientists must be good problem-solvers. This is true, but the key to solving problems is not your problem-solving skill; instead it is your skill in posing the right questions.

Often when you hit upon the right question – a particularly pertinent one – you almost immediately have solved the problem.

Asking questions is central for a related reason, one that might seem vague, even mystical. Humans have great creative power. The process by which this power is released involves several steps. First we think about a certain issue. After having formed our thoughts, we translate them into words by either saying them aloud or writing them down. Ultimately this sequence leads to actions that allow us to modify our world. We thus have the following chain of events:

$$\text{Thoughts} \Rightarrow \text{Words} \Rightarrow \text{Actions}$$

This sequence parallels the idea succinctly stated by Emerson (1841): "*An action is the perfection and publication of thought*" and formulated by Saint-Exupéry (1931) that "*in life there are no solutions. There are only forces operating: you have to create these forces, and the solutions follow.*" Our thoughts, made more tangible by putting them in words, are crucial to giving us direction and moving us to action. Our thoughts are ethereal. This is not to discount the value of pictures in our minds. The ideas of some of the more creative scientists grow in their minds as pictures rather than words.

> *The words or the language, as they are written or spoken, do not seem to play any role in my mechanism of thought. The psychical entities which seem to serve as elements in thought are certain signs and more or less clear images, which can be 'voluntarily' reproduced and combined.*
>
> Albert Einstein, in Hadamard, 1954

Words nevertheless are the vehicle to capture these thoughts and make them more tangible. These words can be written, spoken, or simply repeated and quietly pondered in our mind – or any combination of these. Words are more tangible than thoughts and more connected to matter in the case of spoken words (sound waves) or written text. It is through our actions that we translate the words into material events.

There is a simple technique that you can use to generate a large number of questions about a research topic. Take a pencil and blank sheet of paper, and write down any question that you can think of about the research topic. Again, don't filter your questions, but freely associate and write down everything that comes to your mind. Do this for an hour and then stop. As with any effort at concentration, it is best to do this in an environment where you are alone and undisturbed because any distraction can interrupt the flow of questions.

This process is most effective when using free association, writing down any and every question that comes to your mind. Later you will discover that many of the questions you have written come across as being "stupid." That is perfectly fine. You can always throw out these questions at a later stage. If, in contrast, you are too critical of your stream of questions at an early stage, your mind might do so much filtering that you fail to write down potentially "good" questions as well. The quotation marks used here emphasize the judgmental and therefore subjective nature of the words *stupid* and *good*. It is difficult to assess at the outset what is a good question. What at first appears to be a "dumb" one, can often turn out to be a winner, one that is crucial for the direction of the work that follows. Therefore, don't be too critical in the process of writing down the questions that come to mind; write them all down.

Note that, in the above, we have suggested that you write down these questions, starting with a blank piece of paper rather than follow the alternative up-to-date approach of typing them at a computer. Computers can be distracting for this process: Email begs to be read, the internet must be surfed, and a host of projects demand to be addressed. More important, you won't find any questions in a computer (Scales and Snieder, 1999). Starting with a blank sheet of paper forces you to focus on the single task of coming up with and writing down questions. Another advantage of using paper is that later you can cut the paper into little slips that contain one question each, allowing you to easily shuffle and order your questions at a later stage. You will likely find it advantageous at some stage to transfer your questions to a computer, and certainly the word processor can be handy for doing the shuffling as well. You might, however, be surprised to find that hand-sorting of slips of paper offers a particularly effective, free-form means of rearranging your questions while also allowing your thoughts to roam over ideas and further questions inspired by your initial list. Of course, you are free to use the tool – paper or computer – that works best for you. There are no strict rules for doing research; why not try out some options and find out what works best for you?

The importance of generating ideas and questions while giving free rein to association was mentioned by Freud (1899), who quotes the following from a letter written in 1788 by the poet–philosopher Schiller:

> It is harmful for the creative work of the mind if the intelligence inspects too closely the ideas already pouring in, as it were, at the gates. Regarded by itself, an idea may be very trifling or very adventurous, but it perhaps becomes important on account of the one that follows it; perhaps in a certain connection with others, which may seem equally absurd, it is capable of forming a very useful construction. The intelligence cannot judge all these things if it does not hold them steadily enough to see them in connection with others. In the case of a creative mind, however, the intelligence has withdrawn its watchers from the gates, the ideas rush in pell-mell, and it is only then that the great heap is looked over and critically examined.

It is worthwhile to give each sentence in this quote time to sink in and to ponder the application to research.[1] Some of Schiller's statements might appear to be outrageous for a scientist whose work is thought to be based on logic. Schiller states, for example, that "the intelligence has withdrawn its watchers from the gates," as if intelligence prevents us from being creative. How preposterous! And yet he might be right. We argued in Section 2.3 that, even though

---

[1] You might consider the implications of this quote for other parts of your life as well.

logic underpins science, the path taken in research often departs from being logical. At times we can be most effective in finding the path to solution by creative thinking driven by intuition rather than by pure logic:

> ...the temporary relinquishing of conscious controls liberates the mind from certain constraints which are necessary to maintain the disciplined routines of thoughts but may become an impediment to the creative leap...
>
> Koestler, 1964

This is not to say that intelligence is not essential to research. Of course it is. In addition to their value in logical pursuits, intelligence and expertise are there to steer free association at a subconscious level.

The process described in the quote above is nowadays referred to as *brainstorming*. While we often think of brainstorming as a group activity, it can actually be highly productive when done individually. In true brainstorming, the ideas come in by free association. Unfortunately, it too often happens that, when brainstorming with a group, the temptation is large to react on each other's ideas in a way that is destructive for advancing the brainstorming process. Comments such as "I don't think that is right" or "that can never be done because..." interrupt the flow of the free associations needed for productive brainstorming in a group.[2] In our experience, brainstorming sessions usually start on the right track with a group process of free association, but, too often, after a while this process becomes stifled by critical comments from within the group. The later into the session to which such comments can be postponed, the more useful will be the brainstorming. This is not to denigrate the value of brainstorming with others. In the best of group brainstorming, ideas abound, but an effective chairperson is needed to monitor or facilitate the process by redirecting the flow of the discussion when participants become reactive.

---

[2] Such interruptions to the free flow of thought exist not just in groups; they can break into your individual brainstorming as well.

Questions, of course, can arise in settings other than brainstorming sessions of the group or individual. Questions pop up while reading a paper, listening to a seminar, or at random moments during the day or night. To ensure that these questions are not lost, keep pencil and paper handy so that you can write them down immediately. This can also be done in the margin of whatever (of your own) you are reading or on index cards kept close by. It is useful to carry a small "questions" notebook, suitably labeled with a big question mark on the cover. Alternatively, a handheld computer can be used for this purpose, although the authors do not see themselves operating such a device in the middle of the night after a particularly inspiring dream.

5.2 ORDER AND PRIORITIZE QUESTIONS

When there is no problem, there is no incentive to think about it or solve it. Only when a problem becomes irritating enough do we tackle its solution.

Killefer, 1969

The list of questions made by free association and by collecting questions over time basically frames the problem or problems that you face in your research and that prompt you to take action. Usually, embedded in the questions you have formulated is a logical order, perhaps initially hidden. Because research is most effective when its actions are well ordered, once you have made a list of questions about the topic of your research it is therefore appropriate to organize them into some logical arrangement. As mentioned above, this can be done particularly simply by cutting the sheets of questions in pieces with one question per paper slip. These smaller slips of paper can then be moved around the table until they reflect a logical order. Prioritization of the different questions can be accomplished by defining categories for those that are related to one another and then sorting the slips of paper within each category. The shuffling of questions can, of course, also be carried on a computer since text can be placed at will on the screen.

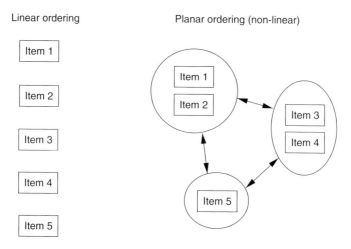

FIG. 5.1. A linear ordering of items (left) versus an ordering of items in a plane (right) that allows for more creativity and flexibility.

Shuffling questions around on a sheet of paper or a computer screen might appear to be a primitive way to organize questions, but it has an important advantage over making a single sequential list. Having items listed simply in sequential order draws the individual into thinking linearly. Such an ordering is shown in the left part of Fig. 5.1. In contrast, shuffling small pieces of paper around over a plane, as shown in the right part of Fig. 5.1, offers freedom to make connections among the different items more readily than with the linear ordering in a conventional list. Once you paste the pieces of paper onto a large sheet of paper, you can use colored markers to draw arrows that show connections among different items, or you can write additional comments on the master sheet of questions that you have created. Alternatively, the drawing tools on a computer can be used to achieve the same results.[3]

You might think that we are describing a merely kindergarten approach to creating order, but consider the method Mendeleev used

---

[3] The shuffling, either of paper or on the computer, has yet another advantage. Our thought processes do not shut off during the shuffling. Quite the opposite; expect the shuffling process to generate yet more questions.

FIG. 5.2. The periodic table of elements as compiled by Mendeleev in 1869. The table is rotated 90 degrees compared to the way it is currently displayed.

in the nineteenth century, which led him to discover the periodic table of elements. An early version of his periodic table is shown in Fig. 5.2. As he describes,[4]

---

[4] Mendeleev, *Principles of Chemistry*, Vol II, 1905.

I began to look about and write down the elements with their atomic weights and typical properties, analogous elements and like atomic weights on separate cards, and this soon convinced me that the properties of elements are in periodic dependence upon their atomic weights.

By making a card for each element, and by ordering them according to properties of the various elements, he discovered an underlying order to the properties of the elements. Mendeleev accomplished this at a time when about 50 elements that we know today had not yet even been discovered! It is a tribute to his genius (and doggedness) that he perceived the order in the periodic table when so many pieces of the puzzle were not yet known. The early draft of the periodic table in Fig. 5.2, which clearly shows that the intellectual work is still in progress, is a two-dimensional display of his ideas much in a fashion akin to shifting pieces of paper.

The process of generating questions and ordering them is applicable throughout your research career. One of us, together with a graduate student, visited Shell research with the goal of giving shape to a joint research project. We spent the first two days talking about possible research directions, about things we knew and those we did not know. In the process we generated question after question, and wrote down every question and idea that arose in the discussions. On the second evening in our hotel room we borrowed scissors, glue, and a large piece of paper from the surprised hotel staff, and cut our list of questions into pieces with one question on each piece of paper. We ordered these questions into a coherent and logical pattern on a large sheet of paper. The next morning we could show our colleagues at Shell a workplan for the project that needed little modification from their side.[5] Quite likely you might consider it unecessary to revert

---

[5] This was, in fact, not the end of the story. The student lost the research plan while going through security at the airport in Houston. We could not retrieve the research plan from the security company, but to our surprise we could reconstruct

to such kindergarten steps as cutting, shuffling, pasting, and drawing colored arrows. Give it a try though, to find out if this approach works for you. A variation of it did for Mendeleev.

5.3 TURNING QUESTIONS INTO A WORKPLAN

We've just discussed making a prioritized list of questions that support the primary question of the research project. This list serves a two-fold goal. First, it helps to focus on the issues that you want to investigate. Second, the list can be used to translate those questions into actions. Sometimes it is difficult to decide where to begin when addressing a problem. Your prioritized list, however, can be a valuable aid in deciding the order in which to proceed in carrying out your investigation. This amounts to making a workplan for the research. Such a workplan should not be viewed as a formality for documenting the work to be undertaken. Instead, making a workplan is a process of formulating a strategy for carrying out the research in a systematic way. Developing and following a workplan is critically important when you are working alone, but is even more so when you are working in a group. To use the full potential of the group requires cooperation, dividing of tasks, and coordination of actions. While working in a group poses problems associated with interactions among members, potentially a group can make much more progress than can an individual working alone. Whether you work alone or in a group, the workplan is an important tool for giving focus and direction to research (see also Chapter 6).

A good workplan starts with a clearly defined goal of the project. Write the overall goal at the top of the workplan, and keep it in mind throughout. In a sound workplan, all the actions in the plan ultimately support the goal of the project. Having the plan written down and prominently displayed aids in keeping focus on the project's goal.

the research plan within an hour. Once the research plan had been internalized (i.e., imprinted on our mind – in no small part through the manipulation of the scraps of paper), the physical manifestation of the plan in print diminished in importance.

The task of making a workplan is eased by starting with the ordered list of questions, discussed above, and then translating the questions into actions that could help solve them. When working in a group, selected activities can be assigned to different individuals, or to subgroups. A good workplan contains the following elements:

- An ordered list of activities to be carried out.
- A clear indication of how these activities are interrelated.
- A deliverable and a timeline for each activity.

A deliverable can be, for example, a written or oral report, a set of measurements, or entries for a database of references in literature search. It is essential that the workplan be documented. We have good experience with documenting the workplan in *html* format, the format used for webpages.[6] In this way the workplan can be read on any computer that has a web browser. An additional advantage is that the workplan can also be posted on the internet or intranet – a version of the internet that allows access to only certain users. A *Wiki* is particularly convenient for sharing information with a selected group through the internet. A Wiki is a webpage that can be accessed and edited conveniently through the internet by a designated group of users. The web-based tool to edit a Wiki is user-friendly, and it is easy to set up such webpages.[7]

When working in a group, it is important to hold each other to the workplan and its timeline. Suppose for example that, in a group project, a subgroup is scheduled to report to the full group on a certain issue at the end of the month. It remains a good idea to go ahead with the meeting to discuss the subgroup's part of the project at the scheduled time even if that subgroup otherwise might not feel ready to give its report just then. Useful types of questions that might arise in this

---

[6] Numerous editors are available for providing output in html format; hence, there is no need to learn and apply the tiresome rules of this format.

[7] For more information see http://www.wiki.org.

circumstance could include the following. What are the factors that cause the subgroup to feel unready to report? Were there unforeseen complications that bear discussion within the full group? When will the subgroup be ready to report? Is it necessary to change the workplan? Even when you are working alone rather than in a group, posing such questions can be a useful exercise because it helps keep your work on target.

While a workplan is a useful tool for keeping focus in research, it is nothing more than a tool to serve the goal of carrying out the research. The workplan itself is not the goal and should never be turned into one. Moreover, the workplan should never be considered as carved in stone. It is in the nature of research that its course can rarely be foreseen in all its details. If it could be, the research probably is not sufficiently innovative. It can therefore be expected that the workplan will need to be modified, perhaps frequently, over the course of the project. Don't be hesitant to adjust the workplan as the project progresses. Again, it is nothing more than a tool to formulate a strategy. Hanging on too rigidly to a single strategy could hamper progress in the research.

Although this chapter opened with an example of questions one could think of about a subject that is outside the individual's field of expertise, for the research that you'll be doing your questions ought not to be arriving totally out of the blue. Rather, they will come from a field of research, even if it is new to you, about which you have become quite familiar. Key to gaining that familiarity is doing a thorough literature search. Such a search is the start of any research project, and it can be worthwhile periodically to repeat the search while carrying out the project. Chapter 9 offers suggestions for aids to making that search effective.

The central message of this chapter is that, whatever background you bring to a problem, *your quality as a researcher depends primarily on your ability to ask the right questions, but that can happen only if you pose lots of questions, many of which will subsequently be discarded.* In case you have doubts about this approach, watch leaders in a research field during a seminar. These usually are the individuals who come up with numbers of questions. Their ability to generate questions, the product of an open mind, is no small factor in what makes them leaders in their field.

6   Giving direction to our work

6.1 SET GOALS

If you don't know where you are going, any road will take you there.

Lewis Carroll, 1832–1898

It's a dream until you write it down; then it's a goal.

Simon, 1998

It's difficult to imagine embarking on a journey, adventure, activity – any endeavor – without having a goal, however vague that goal might be. Even if you don't study a map before going on a road trip, you at least think to put gas in the car. Goals for a holiday might be explicit or implicit, and they can range from short term to intermediate and somewhat long. A career in science, starting from your period in graduate school and continuing into a life of research, is a journey, a long one. Much more so than for a holiday journey, the thoughtful setting of explicit goals is of crucial importance for a successful career in research and for success in the research itself. By *success*, we mean here the achievement of valuable contributions in your field, accomplished with a good deal of pleasure and a minimum of needless pain and time wasted.

Goals give direction to our actions. By clearly choosing and defining goals, we provide a focus for action needed to arrive at a hoped-for destination or outcome. Defining goals not only helps in creating a mental commitment to take certain action, it also enables us to formulate a plan of attack toward reaching the desired ends. The most short term of goals, for example, what you plan to accomplish today, are the easiest to define and pursue. The focus needed for success in graduate school and in research in general, however, requires the defining of goals that are of significantly longer term. With well-defined, long-term goals spelled out, you can then proceed to define a sequence of progressively shorter-term ones. To be clear about our meaning, a well-defined goal minimally requires that it be stated in writing.

For good reason, you can expect your goals to change with time. With a sequence of goals kept clearly in mind, however, you can be in control of those changes and avoid being thoughtlessly deviated from your path. Numerous factors can deviate us from our goals or our planned timetable. In research, setbacks are common: experimental equipment might not be cooperative, computers crash, people might try to convince us to choose a different path, and the general level of distraction can be so large that staying focused is difficult. Moreover, the results of our research can fall short of what we might have expected or hoped for. For these and other reasons it is not always easy – and sometimes it is not even desirable – to keep working toward the realization of a defined goal. Nevertheless, without having set well-defined goals, the risk is high that chance events (and, sometimes, other people) will determine where you are headed. It is unlikely that you would be happy with that. That being the case, *you* need to be the one setting, following through on, and, when appropriate, modifying goals for your research and career.

In research it is likewise crucial to define goals. What do you want to discover?[1] What do you want to achieve? Satisfy your scientific curiosity? Publish many papers and advance rapidly in your career? Develop a patent or marketable product? The attainment of such general goals requires that more specific ones be set and addressed beforehand. Your goal could be to find out how a specific biological process works; it might be a desire to make your name by presenting a breakthrough at conference X. Decide on what

---

[1] In his graduate class "The Art of Science," one of the authors was somewhat astounded to find that so many of his students could not readily complete the sentence, "The goal of my work is..." Likewise, many students had difficulty in completing the sentence, "The questions I'd like to address in my work are..."

you *really* want to achieve, and then work hard to reach your goal. This is the only way to maintain control over the direction in which you head.

Appropriate for research is to set a hierarchy of different goals, some of them global and long-term, which require the defining of shorter-term ones that must be satisfied along the way. An example is "I want to develop a career in research with the aim of solving a grand-challenge problem in genetic engineering, and later in my career I want to share the expertise I have acquired with a younger generation." Working toward such a goal requires study in graduate school and the setting of shorter-term goals for graduate studies such as those for course work and research. The satisfying of shorter-term ones amounts to having passed milestones along the road. A milestone can be, for example, the completion of a certain set of experiments, the finishing of a certain amount of course work, or the presentation of a certain piece of research at a scientific conference. In practice, we benefit not only from defining our intermediate goals but also from monitoring our progress over time. Defining the milestones therefore entails not only *what* exactly is achieved, but also *when* each milestone is reached.

Not everybody has clearly defined goals for the longest time-scales, and, to be realistic regarding these longer-term ones, we should be alert to "expect the unexpected"; unanticipated events in life often give the realization of long-term goals an unexpected twist.

We find the question "what defines success?" to be particularly useful when defining goals. Answering this question necessitates that we thoughtfully articulate just what is our larger goal and, by inference, what are the milestones along the way toward that goal. The question of what defines success can be addressed for each of these milestones.

In research, one usually proceeds from one phase to the next. A phase in research is usually focused on answering a specific question or range of questions. This could, for example, involve a literature search, the formulation of a research plan, the creation of software, or the performing of an experiment. Often, it is essential to finish each step, defined by a specific milestone, before moving on. One might need the results of a specific task in order before proceeding to the next. For example, the creation of a workplan depends on a thorough knowledge of the current state of research in the field, so a literature search is necessary in order to learn what is presently known. The design of future experiments likely will depend on the outcome of measurements previously made. In practice, we work toward our goals by realizing a planned sequence of subgoals (milestones). Confusing the order of the milestones can lead to an ineffective line of research. With a well-thought-through sequence, reaching a particular milestone gives confidence to take the next step in research.

One model for defining goals in research is to structure the plans as if preparing a research paper. Such a paper usually begins with an overview of the current state of research and previous work. Next, the research methodology is introduced and then applied to experiment, numerical simulations, or to data that are collected otherwise. This information is then processed and applied toward the research question being asked. Finally, this is all integrated into conclusions and a discussion of the implications of the work done. This generic structure for scientific papers can effectively be used for defining goals and milestones for research. In practice, one does not know ahead of time just which part of the research will be successful, so it is prudent to lay out a number of potential paths that could be followed.

Again, in defining goals, we are choosing where we are headed. Then, by articulating our goals, sharing them with others, and writing them down for ourselves, we build a commitment to work toward realizing them. This commitment then leads to actions that bring us closer to reaching our goals. Note that, in doing so, we follow the same chain of events as discussed in Chapter 5: thoughts (our goals) lead to words (the articulation of goals) and then to actions (aimed at realizing goals). Following this sequence helps us use our creative power to give shape to our future.

6.2 FIVE STEPS TO TAKE IN WORKING TOWARD YOUR GOALS

Intentness is the ability to resist temptation and stay the course, to concentrate on your objective with determination and resolve. Impatience is wanting too much too soon. Intentness does not involve wanting something, intentness involves doing something.

Emery, 2008*

As we have seen in Section 6.1, without having defined your goals, other people and external circumstances – including chance – will decide where you are heading. Defining goals, however, is a different matter from reaching them. A process of translating goals into actions might be the following, adapted from the book of Robbins (1997).

- *Define your goals.* Again, being explicit about your goals, making them sufficiently specific, and writing them down helps you create a strong commitment to them so that they can be translated into actions. Remember the question "what defines success for my project?" Answering this question often leads to a more clearly specified, and sometimes more measurable, endpoint of the chosen goals, which can help you work toward their realization. Note also that having a timeline will be particularly helpful once you've defined your goals.
- *Decide if you are willing to pay the price.* Achieving something large in life rarely comes for free. Just as becoming a great athlete requires training, a shining academic career requires dedicated study to learn both your chosen field and the professional skills needed to be successful. This will entail hard work and, at times, some sacrifice. You should make a conscious choice as to whether or not you are willing to follow through on your larger goals. To excel in research you must be willing to – that is, make the conscious choice to – put in the energy necessary to learn new things, follow the scientific literature, establish contacts in the scientific community, take steps

* Reprinted with permission from Firehouse Magazine, Copyright April 2008.

to foster your creativity, and have the courage to be truly innovative. A mismatch between your goals and your willingness to invest in them is a recipe for disappointment. Your choice can work out either way, *but choose*!

- *Define a strategy.* Once you have decided that you truly want to realize your goals, define with care your strategy for achieving them, for planning steps to take, and for defining milestones along the way toward the goal – and don't forget to include a timeline. Again, put your strategy on paper in order to make it more tangible. Then start executing the plan.

- *Evaluate your strategy.* While executing your plan, you will probably discover that some portions work out well, and others don't. This is normal; nobody is perfect, and no plan is fool-proof. This is aggravated by the fact that "in the beginning we don't know what we don't know." Our initial ignorance often leads to a strategy that needs adjusting. Therefore, periodically evaluate your strategy by identifying its weak and strong points, assessing what works and what does not.

- *Modify your strategy.* Adjust your strategy based on evaluation of which of your actions have been effective and which have not. Likely you will go through cycles of re-evaluating and modifying your strategy, making changes to the execution of your plan.

Neither setting nor following through on goals is a trivial matter, but realizing that you need to modify your goals and strategy, and having the discipline to define and act on the changes, can be the hardest part. These steps are applicable to any type of large goal in life, including, of course, your research goals. You might find it instructive to re-read the above sequence of steps with your research project in mind.

6.3 WHAT IS YOUR GREATEST RESOURCE?

In following the above steps of defining your strategy, evaluating your plan, and adapting the strategy, you are not alone. You are surrounded by people who can help in a variety of ways. A good way to use the experience of others is simply to observe what they are doing. Suppose you are struggling with an academic course you are taking, and this course appears to be much easier for a friend. Much as your inclination might be to trudge through the material by yourself, it can be worthwhile to see how your friend learns the course material and to compare her approach with your own strategy. It could simply be that your friend learns certain types of material more easily, but often it could be that her study habits are more effective than yours. But, you can do much more than just observe. Why not ask colleagues how they approach subjects or problems, and even see if you can study alongside them? Most people are willing to help and share their expertise; certainly that is true of friends.

Consider two basic ways of learning. The first is to learn by encountering your own mistakes and then figuring out how to correct them. The second is to imitate the methods of others. Although both approaches to learning, or a combination of both, work well, the second one, emulating others, can be a much more efficient investment of your time. The essential point here is that you are likely to be surrounded by a wealth of expertise and possibilities for getting help. *The people around you can be your most valuable resource.* Most people are flattered for you to have sought their advice, including for help in defining, evaluating, and adapting your strategy on the basis of the evaluation.

A corollary of the principle that your colleagues can be your greatest resource is that you can be a great resource for them as well. Not only do you learn from others, but others learn from you, often without actually being aware of your being the teacher. We learn much from simply observing others; likewise we often influence others through just our words or actions.[2] Whether casually through our actions or explicitly through study together, a network of colleagues

---

[2] This is one reason that "leading by example" is such a powerful tool in management. Whether we are aware of it or not, we are being observed by those around us. A discrepancy between the words and actions of a leader rapidly undermines his or her credibility.

who mutually teach and learn is a wonderful and powerful way to advance expertise.[3]

6.4 BEING GOAL-ORIENTED OR PROCESS-ORIENTED

While goal setting has the value of forcing us to focus our actions on predefined results, the emphasis on formulating goals and working toward them carries a certain danger for some people. We are all different, and we all have different driving forces that guide our behavior. For some people, the primary desire is to reach their goals no matter how demanding and difficult they are to achieve. For them, the satisfaction of having reached a goal can be enough to endure months, or even years, of arduous work that might not be intrinsically satisfying. Other people are much more focused on the *process* of working toward their goals.

As an analogy, consider mountain climbing. For some climbers reaching the summit is the sole reason for climbing a mountain, despite the fact that the process of getting there brings hardship, suffering, and even physical dangers. Other climbers experience the joy of simply going up the mountain. They enjoy the views, the teamwork involved in a climbing trip, and the physical exercise. Even though for these climbers the summit is a goal, they are not primarily driven by that goal, but by the enjoyment of the process, the climb itself.

There is, of course, no hard distinction between those who are predominantly goal-oriented and those who are mainly process-oriented. It is worthwhile, nonetheless, to be aware of these differing motivations. In the reality of everyday life we do not always achieve our goals; especially in competitive situations. Being too much oriented toward goals, makes a person vulnerable to disappointment and frustration when a goal is not reached. Furthermore, the moment of reaching a goal is usually short-lived, while the process of working

---

[3] The graduate experience is much more than what you learn in class and from research that you do with your professors. An invaluable part of your education derives from what you can learn from your fellow students and what you find that you can teach them.

toward a goal is often lengthy. For this reason, it is valuable not to focus exclusively on reaching goals, but also to enjoy the journey toward that goal.

The following is modified from an article written by Portia Masterson, who owned a bicycle store in Golden, Colorado. In her newsletter *Self Propulsion*, she describes her satisfaction in tuning a bicycle, an example of the enjoyment we can get from being absorbed by doing something we simply love doing.

> When I focus on the process [of tuning bicycles], the step-by-step progress of each task and pleasure, I do my best, and, as a result, enjoy it all the more. I clearly define the end result I am seeking and then settle into details. If my mind wanders off to other responsibilities or recreational plans, I patiently bring it back to the process. . . .
>
> On days when I would repeatedly drop tools or make several mistakes in a row, I used to despair and resign myself to a day of blundering. Now I know it is my choice, so I laugh at myself and take control. I tell myself that the choice is mine and I can turn this trend around simply by focusing cheerfully on the process and letting go of the pressures or concerns that are distracting me. . . .
>
> Not only does attention to process improve my patience, it makes the job interesting. I don't get bored; each task has variations. Even after truing hundreds of wheels, each truing is different and there are things to be learned.
>
> At the end of each day it is equally vital to process the details of the day. What was said, done, and experienced? When the day was pleasant, I can learn from what I did well and try to recreate that again. When the day was challenging or exhausting, I can sort out what contributed to my difficulties and try to avoid them in the future. When I process the events and attitudes, I am sometimes surprised to discover I created the entire mess in my head. Whew! Then I can let go of my poor mood and move on.

So try giving process your attention and see if the details of life aren't more fun and more rewarding. Perhaps processing your perspective will lead to a better outlook or the resolve to change distressing circumstances.

Masterson, 2002, unpublished document, reprinted with permission

Many aspects of this description of Portia Masterson's experience of process in the repair of bicycles can be translated to those of doing research as well. Consider the example that she gives of dropping tools at moments when nothing seems to work. As a researcher you might have a similar experience; on some days the computer program that you write is particularly full of bugs, laboratory tests fail, or none of the equipment that you are building seems to work. At those moments you can benefit from stopping the worrying and hurrying, and instead try to get caught up into the process of being absorbed in a task that you want to do well. You might go through the above quote again and seek analogies with your own research experience.

At moments when you are completely absorbed in your task, just as for Portia Masterson, any sense of time disappears because all attention is focused – without explicitly realizing it – on carrying out the task as well as possible. The psychologist Csikszentmihalyi (1991) calls this the "Flow-experience," which, for different people, occurs when involved in various activities. You might recognize it from your own experience while making music, playing sports, repairing cars, pursuing spiritual or religious activities, or doing research. Csikszentmihalyi makes the point not only that we are at our best while experiencing *Flow*, but also that spending an insufficient amount of time in the flow-state, leaves us unhappy and unsatisfied.

We have argued that not only is it essential to define carefully and focus on goals, but we are at our best when we find enjoyment in the process of working toward those goals. Being goal-oriented and process-oriented are not necessarily contradictory. The key is to strike a healthy balance between our focus on the goals themselves and the process of working toward them. The beauty is that being absorbed in the process of reaching a goal often makes us more effective in actually realizing that goal. Of course, not every task along the way to meeting a goal will be inherently joyful. Nevertheless, the effort to consciously seek joy in whatever we do not only can help make the task less onerous and even more pleasurable, it can offer the best prospect of yielding a satisfying outcome.

6.5 MORE THAN GOALS AND PROCESS
IS *MEANING* IN OUR WORK

More than the goals that we set and the satisfaction that we can achieve from the process of our work, that work might or might not have special meaning and value to us or others. The concept of meaning is much less tangible than that of goals, but this makes it no less important. The meaning of work is described by Chittister (1991) in the following way:*

> ... we must learn to distinguish between purpose and meaning in life ... Purpose has something to do with being productive and setting goals and knowing what needs to be done and doing it. It is easy to have purpose. To write seven letters today, to wax the floor, to finish this legal brief, to make out those reports, to complete this degree, that's purpose. Meaning on the other hand, depends on asking myself who will care and who will profit and who will be touched and who will be forgotten or hurt or affected by my doing these things. Purpose determines what I will do with this part of my life. Meaning demands to know why I'm doing it and with what global results.

Difficult as it is to set goals and to have a purpose, these are easier than giving shape to the meaning of our efforts. The meaning of anyone's work is personal and individual; a given activity can have different meaning for different persons. This meaning can also depend

* Copyright © by Joan D. Chittister. Reprinted by permission of HarperCollins Publishers.

on the time and place of the work, and very much also on our outlook in life. Chittister (1991) stresses that meaning as well depends on how our work affects others, either positively or negatively. It moreover depends on our connections with others; our endeavors are not carried out in isolation from others.

We can offer no simple recipe for defining and realizing the meaning of our efforts, yet the meaning attached to our work can ultimately be the aspect most important to success in our career. This measure of success does not depend on the number of papers that we publish, on our income, or on how high we rise on the hierarchical ladder. It depends primarily on how we have touched the lives of others. A career in research or education offers numerous opportunities for this.

7 Turning challenges into opportunities

The brick walls are there for a reason. They're not there to keep us out.
The brick wallls are there to give us a chance to show how badly we
want something.

Pausch, 2008*

Seldom is the path toward success in research either straight or fully charted in advance. Stumbling blocks abound in any research. In this chapter we cover some of the common ones and offer suggestions for steps in conducting research aimed at minimizing their harm and even turning them to advantage.

7.1 BEING CONFUSED BECAUSE OF LACK OF DIRECTION

Research can be a confusing activity. Your research plan might be poorly formulated; worse, you might have no plan at all. At times you cannot understand intermediate results of your studies; data that you've recorded might conflict with a theory you developed, or two different lines of reasoning that both seem to make sense give different answers.

The first source of confusion, in which the work basically lacks direction, is clearly a negative one and needs to be fixed early on. Any of a variety of factors can have caused this state of stagnation. Perhaps you have not yet settled on a research topic or you have decided on one but the choice is insufficiently specific to get started on the research. You can avoid this pitfall by being aggressive in choosing and refining a research topic quickly. General considerations when choosing the research topic were given in Chapter 3. Usually, however, the

---

bottleneck is in identifying the right idea, or, when you are a graduate student, in finding the right adviser or research supervisor. In either situation, it is valuable to talk with many different people. Talk with colleagues, with potential research supervisors, with friends, but talk! (In reality, the key of course is not to talk, but to *listen*.) In doing so you are not simply milking people around you for ideas. Rather, talking with others offers a variety of benefits:

- It forces us to formulate our interests and questions. Our thinking is often fuzzy. Formulating our ideas/questions/doubts in words forces us to be clear and specific in our thinking. Perhaps you have had the experience of asking a question to somebody, and halfway through the question you suddenly see the answer. This often leads to the question petering out with the remark "never mind, I see it now," but having voiced the question actually was useful because it has led you to the answer. Apart from the clarity brought by formulating a question, translating your thoughts into words makes them more concrete and tangible, characteristics that help in resolving the question. This process was also described in Section 5.1.
- When thinking, our thoughts often tend to fall into a fixed pattern. Despite its value, too fixed a thought pattern can be limiting as it can put us on a narrow, perhaps erroneous, track. Other people can be helpful in offering the correction needed to move into an alternative, more productive, line of thinking.
- Last, but to be emphasized, talking with others can open an easy route to new ideas. It might be that a colleague directly gives you a new perspective, or perhaps the conversation leads you yourself to the new idea. The mere act of formulating your problem for others can generate fresh ideas in your own mind. Don't underestimate the power of words voiced.

In addition to talking with others, it is important to investigate the state of research in the area that interests you. Read the scientific literature, look through recent conference proceedings, and search the internet. Such searches, as described in detail in Section 9.1, can aid you in developing a focus on a research project.

Even when you have a research project in mind, you could still be confused about the direction in which your research should go. To move forward, it likely is necessary to make your workplan more-specific. In Section 5.3, we have seen how this works. The sequence is to set goals, formulate research questions, and translate these questions into actions – a simple recipe that usually works. When you have trouble translating your questions into actions, these questions are probably insufficiently specific. The appropriate step is to formulate more-specific ones.

7.2 BEING CONFUSED BECAUSE YOU DON'T UNDERSTAND SOMETHING

That awareness of anomaly opens a period in which conceptual categories are adjusted until the initially anomalous has become the anticipated.

Kuhn, 1962*

Suppose the results of your research differ from what you had anticipated, or laboratory measurements or computer simulations conflict with your expectations. Although this confusion can be frustrating, it actually can be a valuable step toward gaining deeper insight into the problem. In carrying out research, we of necessity start with or develop certain expectations. After all, if we had no concept whatsoever of what to expect, we would have no basis for having embarked in that research direction. These expectations are based on our understanding of the research problem at a given time, which, early on, is usually incomplete.

Conflicts between measurements, or other data, and our expectations simply mean that our understanding of the problem is imperfect and needs to be modified. This type of confusion valuably serves as a reality check, forcing us to bring our thoughts in line with reality.

---

* Reprinted with permission of the University of Chicago Press.

Such conflicts can thereby be advantageous starting points for delving into truly interesting aspects of the research. Therefore, *we can profitably allow ourself to be confused at times.* We nevertheless should seek to break out of the confusion as soon as possible and not be stopped by it. The best way to do that is to translate the confusion into specific questions. (You've heard this before.) In the process, dream up wild explanations that might resolve the conflict. Specific questions and off-the-wall explanations might well be the catalyst that leads to a modification of the research plan. In that way you can convert something negative (confusion) into something positive (a better research plan, direction, or concept), with the possible result of making a worthwhile scientific discovery.

Kuhn (1962) argues in his book, *The Structure of Scientific Revolutions*, that radical changes in science are frequently precipitated by confusion in the research community. This confusion, and the associated frustration, causes scientists to seek new avenues in order to make progress. Many of these attempts fail, but the new ideas provide fresh insights that can open up a novel approach to the problem. Kuhn calls a common approach to a scientific problem a *paradigm*, and asserts that science needs new paradigms at times to revitalize the approach to research. A new solution or even a new paradigm is often the welcome outcome of being confused and stuck.

7.3 SIMPLIFY FIRST

A major source of difficulty when starting on research into a new topic can be the complexity of the problem. The problem might be so overwhelmingly complex that it is difficult even to ask meaningful questions, let alone embark on a solution. A good approach in general is to sneak up on the problem: first simplify it by removing its complicating factors, learn from the solution or ideas that follow from that simplification, then add complexity little by little, seeking to gain understanding in increasing depth at each step.

Suppose, for example, that one wants to study the flow of water in an oceanographic problem or describe the flow of blood through veins and arteries. How much detail is needed to account for the dynamics of the water? Clearly, there is no need to take into consideration the interactions among water molecules.[1] Interactions among the constitutive molecules is a complexity that can be stripped from the problem, leaving a description in terms of water as a continuum. At the outset of the oceanographic problem, perhaps salinity and temperature gradients might be ignored; each of these complications can be added in turn as more of the problem is understood. This is rather like the way a new subject is taught to students. The initial version of the problem taught is a simple one aimed at conveying understanding of the general framework. Once that basic understanding is achieved, the teacher can make the topic gradually more realistic by adding in complexity.

Simplifying a problem at the outset reduces the chance of getting lost in details, allowing you to focus on and understand the essential elements of a problem. The examples above are obvious ones in which unnecessary detail can be removed at first, making the resulting problem easier to solve. Usually in forefront research it is not trivial to determine what complicating detail can be removed from the problem in order to make it simpler and yet preserve aspects of importance; it nevertheless typically is fruitful to take this approach. Even if you have over-simplified a problem, resulting in the inability to explain the research question, you will have learned from this over-simplified version that it does not contain the essence of the problem. Then, by adding complexity to the problem, you can progress toward the needed realism. This approach not only makes the research easier, it also leads to a better understanding of the essential elements in the topic of investigation.

It sounds easy to simplify a problem, but in practice this is often not so, especially for beginning researchers. It takes experience to make sound decisions concerning what are the essential elements of a problem and what is unnecessary complication. In order to build up experience, one needs to practice. This can be done, for the problems

---

[1] Perhaps surprisingly, the details of such interaction is still a topic of debate [Bukowski *et al.*, 2007].

encountered, by posing the question "which elements of the problem are truly essential?" You can practice this with your own research, but other occasions offer opportunities for doing this as well. During a seminar you can play the game of simplifying the problem being discussed; "how can your approach be simplified?" is often an excellent question to pose to the speaker. Another way to come to grips with the complexity needed to decipher a research problem is to explain the problem to others in the simplest terms. Formulating the problem in its utmost simplicity for others can give the focus needed to see its essential elements. One of the wonderful aspects of teaching – recognized by most teachers – is that it forces us to deeply understand the material that we teach; the act of explaining material to others usually brings clarity. Another way to understand the essential elements of a problem is to ask for the thoughts of others. Remember, colleagues can be your greatest resource.

Stripping the complexity from a problem can be dangerous when not done with care; the simplification might remove some essential component from the problem. In the study of properties of water other than that of flow, as in the example above, one might indeed have to take into account the interaction of the water molecules in order to fully understand more complicated phenomena such as freezing (phase transitions) and surface tension. In simplifying the problem, you don't want to throw the baby out with the bath-water molecules.

With this warning in mind, a good rule of thumb remains to *simplify your problem first and then bootstrap your way into complexity*. Always keep in mind, however, that while simplifying assumptions can be the key to progress in a problem, you need to remain aware of the constraints that your assumptions have put on the problem, and don't forget to lift them further down the road of your research.

A further caveat is needed for this approach of simplifying a problem and adding complexity only as needed. In Section 2.2, we discussed the distinction between reductionism and wholism. The approach of simplification presented here is natural for problems that are well suited to a reductionist approach. As discussed in Section 2.2, however, this method is less suitable where complexity-of-the-whole forms the essence of the problem. One should be especially cautious about over-simplifying problems that require a wholistic answer.

7.4 IDENTIFY MISTAKES QUICKLY

If you're not prepared to be wrong, you'll never be original.

Ken Robinson[2]

None of us is perfect; we all make mistakes. This is especially true in research, where we are trying to do something that has never been done before. It is perfectly normal to discover at some point that we have erred in some way. Our reasoning might contain a flaw, measurements could have been influenced by some phenomenon that we had not taken into account, the design of a test trial might contain a conceptual error, or we might have missed someone else's important research result.

There is nothing wrong with making mistakes; it is an integral part of doing research – and of life.[3] In fact, the more aggressively you pursue your research, the more mistakes you can expect to make. What is important is that you identify mistakes quickly: unaddressed mistakes cost unwanted delays in your research. Once recognized, however, the mistake itself can be a source of new learning and insight.

It is central to recognize mistakes quickly for yet a different reason. It is natural in doing research to develop an emotional involvement with the research path that you have been following. You become attached to that path and where it has led your research: you develop a certain hypothesis and find yourself wanting the results to have a particular outcome. The deep involvement is a good thing; you need it to be truly passionate about your work. Once you have identified a mistake, however, it is essential to let go of some of your ideas. Unfortunately, with the emotional involvement that you have

---

[2] From his TED video lecture "Do schools kill originality?". http://www.ted.com/index.php/talks/view/id/66.

[3] As an aside related to the value of doing careful search of the literature, you don't want to make *old* mistakes. It's best when your mistakes are new ones.

developed, this can be difficult. The longer you have worked in a faulty direction, the harder it becomes to acknowledge your error and change the direction of research. Therefore, once a mistake is discovered, correct it as quickly as possible. Painful as it can often seem, you need to drop your attachment to an erroneous direction, no matter how much time and effort you have invested in it.

As we indicated earlier, errors are not necessarily harmful. A classic example is the serendipitous discovery of penicillin by Alexander Fleming in 1928. Fleming investigated the growth of staphylococci, a bacteria that is a common source of infections in humans. He was growing the bacteria in culture dishes. One of these dishes became contaminated by a fungus, and Fleming noted that the staphylococci did not grow in the vicinity of the fungus. From this observation, he speculated that the fungus produced a chemical that inhibited the growth of the bacteria, an inference that ultimately led to the discovery of penicillin, an antibiotic that has improved the lives of millions. Fleming's failure to ensure clean laboratory procedures that day was a mistake, but one that led to the discovery of an important new drug and that brought Fleming the Nobel Prize for medicine in 1945.

The route along a wrong track can lead to places that, although unexpected, can be highly desirable and would otherwise not have been found. The alternative is to be fixed in a narrow groove that might be headed nowhere. It is when you follow the wrong track for too long that it becomes a serious roadblock. Once discovered, the key to research success is to spot that you have veered from a fruitful direction and then to change course.

7.5 SERENDIPITY AND PLAYFULNESS

The most exciting phrase to hear in science, the one that heralds new discoveries, is not "Eureka!", but "That's funny . . . "

Isaac Asimov

That the course of science is not always predictable can be readily appreciated: At the beginning "you don't know what it is that you don't know." This statement is particularly relevant for research, which aims to discover things we don't yet know. We have stressed the importance of developing a research plan in Section 5.3, but it should be understood that one cannot plan the ultimate course of research in advance. When dealing with the novel and uncertain phenomena that are the subjects of research, guidance such as that offered in this book cannot be taken as a precise recipe for success.

The discovery of radioactivity (Pais, 1986) serves as an illustration. In 1896, the existence of X-rays and knowledge that these rays could penetrate, for example, through black paper was known. In that year, Henri Becquerel was investigating the photoluminescence of X-rays by uranium salts. After exposing these salts to sunlight, he would put them on top of a photographic plate wrapped in black paper and study the exposure of the plate to X-rays from the irradiated salts. One cloudy day, Becquerel decided that the experiment would be pointless to do that day because of the lack of sunlight. He put the uranium salt with the photographic plate wrapped in dark paper away in a cupboard, and soon forgot that he had done so. When he remembered a few days later, he decided for some reason to develop the photographic plate, and, to his amazement, found that the plate was exposed. From this fortuitous sequence, he drew the conclusion that the uranium salt was emitting X-rays even though it hadn't been illuminated by sunlight. The uranium salt carried its own internal source of energy! This realization ultimately led to the discovery of radioactivity.

This story has some curious aspects. First, if the day of the experiment had not been cloudy, the episode would not have occurred. Chance events often play a role in science, and the story of Becquerel is far from being the only discovery where serendipity, "the occurrence of a desirable or beneficial discovery by accident," plays an essential role[4] (Roberts, 1989; Hannan, 2006). But there is a second aspect of this story. Why did Becquerel develop the photographic plate? According to his current understanding, the plate would not

---

[4] Examples of the role of serendipity in science abound. At http://en.wikidpedia.org/wiki/Serendipity are listed some 50 examples of major discoveries in the fields of chemistry, pharmacology, medicine, biology, physics, astronomy, and invention.

have been exposed. Seemingly, it therefore did not make sense to develop a photographic plate that he thought had not been exposed. We don't know why he decided to develop the plate, and can only speculate. Perhaps he had an intuitive inkling of what would happen, perhaps he was not thinking at all (highly unlikely), or perhaps this was just a fun thing to do to satisfy his natural curiosity.

Whatever the reason, Becquerel was not following a rigid research plan based on logic. In a sense, he was playing. When we play, we do something just because we like doing it. Such play is not driven by logic; the activity is driven by intuition and by our desire to do something for the simple reason that we enjoy doing it or perhaps because we are curious about what would be the outcome. The story of Becquerel illustrates how useful a playful attitude in science can be. As we argued in Chapter 2, science often is not strictly the logical activity that many imagine it is. The act of playing opens our mind to possibilities other than what we arrive at through logical process. It also increases the likelihood of chance events that later turn out to be essential in our research.

This is not to say that our research should consist of random playful events. Quoting Louis Pasteur, "In the field of observation, chance favors only the prepared mind." We have stressed the importance of creating a research plan that is driven by questions. Leaving room for play, however, can be useful; it can pay off to be attuned to serendipitous events that are seemingly meaningless at first, but later prove to be essential to making progress.

7.6 BEING STUCK

How wonderful that we have met with a paradox. Now we have some hope of making progress.

Niels Bohr, in Moore, 1966

If you are always having success, you are not doing research.

Art Weglein, personal communication

For a variety of reasons, research efforts often can become stymied. Again, as described in Section 7.1, one reason could be the lack of a research plan with sufficient focus. Other reasons might be that there is something you don't understand or that your approach simply does not work. Being stuck can be frustrating, but it is not necessarily negative. Of importance is that you be aware that you have a problem. Just as when you recognize that you've made a mistake or are on the wrong track, once you recognize that you are stuck you can direct your energy to solving your problem. The frustration of being stuck often resolves itself into new and creative solutions.

Being stuck usually means that there is something wrong with your line of reasoning, or that an essential ingredient is missing. Then, you likely need to abandon part of your approach or add a new component to it. Perhaps surprising, this is sometimes accomplished most efficiently by not thinking about the problem at all for a while. Simply stop working on it, and do something else. It happens quite often that we suddenly see the solution to our problem when doing some completely different activity. Putting your problem away does not mean you stop thinking about it. Rather than consciously struggling with it, you start thinking about the problem in a different way by handing it over to your subconscious.[5] While this often opens up a new avenue of thinking, it has the danger that putting your problem away for a while can be too much of an escape from the difficulty. Don't fall in the trap of giving up on problems by ignoring them or by "sleeping on it," but give your struggle a rest periodically.[6]

---

[5] In the book *The Psychology of Invention in the Mathematical Field* (Hadamard, 1954) – a study of the process of innovative discovery in mathematics and physical sciences – the brilliant twentieth-century mathematician, Jacques Hadamard, includes a classic example of discovery after having put the problem away. After being stuck on a difficult problem, Hadamard's friend, Henri Poincaré, another brilliant mathematician, arrived at the solution seemingly from out of the blue while stepping off a bus during a holiday. Based on this and other similar experiences, Poincaré came to the belief that it is the *subconscious* that works out the solution to a problem, but only after having been "assigned" by the conscious to work on the problem. That assignment came from dedicated conscious struggle with the problem.

[6] Hadamard (1954), quotes Helmholtz who says that "happy ideas" (his word for illumination) never come when his mind is fatigued or when he is seated at his work table, "...there must come an hour of complete physical freshness before the good ideas arrive."

In any research that we do, we work with premises or models. We need these premises in order to make progress, but often are unaware of them, or simply lose sight of our premises, assumptions, and implied models. As you are likely aware, for example, most of the natural sciences rely on the fact that nature behaves according to the laws of mathematics. This "fact" actually is an assumption, a deeply embedded one. Yet many scientists never thought about this far-from-trivial issue, and most educators never even mention this assumption. Nobel Laureate Wigner (1960) gives an interesting account of what he calls "the unreasonable effectiveness of mathematics in the natural sciences." There is no fully supportable reason why nature should behave in accordance with the mathematics that we use. We nevertheless use this principle as an unproven cornerstone of the physical sciences. It is an example of an assumption that many of us use on a daily basis, without even being aware of it.

Mathematicians call such an unproven premise an *axioma*, in religion it would be called a *dogma*, and in the physical sciences the phrase *paradigm* is used. Note, again, that most scientists never give much thought to questioning the fundamental premises of their field. Where it is necessary to proceed from unproven basic assumptions, these assumptions can be the Achilles heel of our research if we are not aware of them and they turn out to be inaccurate. According to Kuhn (1962), scientific revolutions are usually initiated by a change in some paradigm that underlies the science.

In both science and life, we often get stuck because some of our premises are wrong. Recently, one of us was looking for a book on his bookshelves and could not find it. He knew the title and the author of the book, and recalled that it had a black cover – or so he thought. After a frustrating search, repetitively going through his bookshelves, he gave up and started thinking about who could have borrowed the book. Later that day, he felt compelled to try looking for it again and found it within 10 seconds. The book had been staring him in the face during his earlier search, but the cover was blue instead of black. The hidden and erroneous premise that the book had a black cover prevented him from finding the book. The lesson here when you are stuck and confused: *check your premises.*

This advice might sound easy to follow, but often it is difficult to be fully aware of all the premises that underlie our work, especially when they are part of notions that have become familiar to us over a long time. It is useful to create a list of assumptions, the obvious ones and those that are so deeply embedded as to be implicit; ferret them out, write them down, and review them periodically. In this process it can be profitable to use free association, as described in Section 5.1. Not filtering the assumptions as you uncover them and write them down offers the best opportunity to develop a rich and exhaustive list. Once you have that list, you can order it in the same way as when developing the research plan (Sections 5.2 and 5.3). You can then go through this list and verify whether the assumptions are valid or not. This validation or falsification might require further research and stir more questions. Again, as argued by Kuhn (1962), overthrowing assumptions and thereby changing the paradigm can be our most important contribution to science. This holds not only for the big scientific questions in our field of research, but also for the more modest, detailed problems that cause us to feel stuck in a research project.

7.7 GETTING THE RIGHT ANSWER FOR THE WRONG REASON

And generally let every student of nature take this as a rule, – that whatever his mind seizes and dwells upon with peculiar satisfaction is to be held in suspicion …

Sir Francis Bacon as quoted by Moore, 1993

Sometimes things seem to go really well in research. In a deductive approach we might have a theory and measurements that agree well with this theory, or in an inductive approach we discover a theory that makes sense and explains observations well. In such a situation it is easy to sit back and relax, basking in the satisfaction of a job well done. A dangerous trap, however, lurks in this sense of contentment. Because our research seems to come together nicely, it does not necessarily follow that we are on the right track. It is especially difficult to discover that we are operating under false premises when everything seems to work well.

Key words in the above paragraph are "seems to work well" and a theory that "makes sense" and explains observations "well." Because testing of theory by experiment can support ideas only to some degree of accuracy, an explanation should be deemed adequate to only that demonstrated level of accuracy. A paramount example is Newton's second law of motion, which, for 400 years, had been supported by its success in predicting all manner of observations. It, however, turns out to be "right" only to a level of approximation. That law can be said to be hugely wrong in that it incorrectly describes motion for objects that travel with speeds that are a substantial fraction of the speed of light.

In mathematical derivations and quantitative data analysis, it is not difficult to be right for the wrong reason. A common example almost universally experienced while doing mathematical computation is making two sign errors that, combined, cancel to give the right answer.[7] Another example is data-fitting. Suppose one has collected a set of data points that are being fitted by optimizing free parameters in a theory. The fact that the data can be fit to some degree of accuracy does not mean that the theory is correct. Indeed, in the extreme when the number of free parameters equals the number of data points, the data fit can be perfect, while the theory might be senseless. Mathematical errors, however, are generally much easier to find than are conceptual errors on which we have built our reasoning in the various physical sciences.

Being right for the wrong reason is a mistake that is easily made in science. It presents a particularly dangerous trap in the sciences

---

[7] A delightful account of how to find errors in mathematical work is given by Cipra (2000).

because of the difficulty it creates for discovering that an error even exists – a potential source of delay in progress and, perhaps more serious, of erroneous direction in further lines of research. It is therefore crucial to remain critical of your results and to continuously reconsider not only whether they make sense, but also whether your argument has been accurate and comprehensive.

Karl Popper (1965) pointed out in his book *The Logic of Scientific Discovery* that one can never *prove* a theory in science. In deduction, earlier assumptions could be inadequate, and induction involves a guessing-game concerning the patterns and theory that underly reality. According to Popper, truth cannot be established; one can only falsify a theory, not prove it. Falsification can be demonstrated by using theory to make predictions and then, by making observations (measurements), confront those predictions with reality. A prime reason for Popper's claim that truth cannot be established is that a finding could be right for the wrong reason. In order to avoid this trap, it is necessary not only to confront theory with reality, but even then to remain open to alternative explanations. Success in science requires a great and continued openness to alternative explanations and possibilities. For this, one must have both a determined and open mind.

7.8 KEEP TWO TYPES OF NOTES

In research, you are often immersed in your activities. At that time you know exactly what you are doing and why you are doing it, but often you will be surprised to find it difficult only a few weeks later to reconstruct the details of the work that you have done. In research, a few weeks is a short time. Frequently, a research project requires a year or more of work before you can write a first publication. Sometimes it takes an additional six months before you receive the reviews of a manuscript, whereupon you may have to revise certain aspects of your work. This means that you may need to go back to work you had done almost two years previously. By that time, you probably have forgotten the details of your activities, making it difficult to return to the work to improve it.

The only way to prevent this from happening is to make careful notes. Get a notebook, and make it a habit to make notes of your research activities. These notes can be wide-ranging, including random ideas and questions, derivations of equations, parameter settings of experiments, the outcome of experiments, and names and addresses of people. Develop your own system to keep track of these notes. For example, addresses of people are most efficiently stored in your favorite address database.

Sometimes it is necessary to keep notes for legal reasons. This could happen when research groups are in extreme competition, and when copyright issues are at stake. With notes that are properly archived, they can be used as evidence that certain research is carried out at a specific date. In order for notes to be used for this purpose, they must satisfy a number of criteria:

- The notes must be made in such a way that they are understandable for people outside the research group, because if they are unintelligible they cannot be used as a proof that research has been carried out at a specific time. Don't make the notes too terse, and don't use phrases or abbreviations that are not commonly known in your field of research.
- It is essential that the notes are dated; this makes it possible to establish later at which moment in time a research finding has been made.
- Use a notebook with page numbers printed on each page. In this way it is possible to show that the notebook is complete so that the track record is complete.
- When in doubt as to whether or not research results have been archived with a sufficiently strong legal status, it is of course always possible to have the notes notarized.

We would like to add that, in our careers, we never felt the need to protect research notes in a legal way. For us, the notes served only as a track record of past work. In other fields of research, and in extremely competitive situations, it can be useful to keep the legal aspects of research notes in mind.

The type of notebook described above serves to document scientific work. Apart from keeping a record of these activities, we recommend having a second notebook for writing down questions, ideas, and dream experiments that come to mind. Our moments of inspiration often are brief and are easily forgotten. By having this second notebook nearby, you can write down ideas and insights before they are lost.

You could call this type of notebook the "book of questions" or your "dream-book," possibly suitably marked on the cover with a question mark or other symbol that appeals to your creativity. Use a small book so you can easily carry it with you. The lower the threshold for jotting down questions or ideas, the more likely it is that this book will not only contain a record of your ideas, it can actually be a useful tool for helping give shape to your scientific dreams.

8   Ethics of research

As I embark on my career as a [biomedical] scientist, I willingly pledge that I will represent my scientific profession honorably, that I will conduct my research and my professional life in a manner that is always above reproach, and that I will seek to incorporate the body of ethics and moral principles that constitute scientific integrity into all that I do.

I will strive always to ensure that the results of my research and other scientific activities ultimately benefit humanity and that they cause no harm.

With this affirmation, I pledge to acknowledge and honor the contributions of scientists who have preceded me, to seek truth and the advancement of knowledge in all my work, and to become a worthy role model deserving of respect by those who follow me.

Craig *et al.*, 2003

Just as for all activities in life, research has its principles and standards of conduct necessary to ensure that it be carried out in an honest and honorable manner. Such principles and standards, which may collectively constitute or define the "ethics" of an activity, too often are neither objectively nor well defined. This, however, by no means makes them less important than the Federal and State laws that are used in our society to distinguish between behavior that is acceptable and that which is not. In this chapter, we offer a synopsis of what we consider to be the ethics of research, some examples being quite subtle.

Let us first pause and ask why ethics should at all be an issue in science. We are all aware of the need for, and many violations of, ethics in politics, law, and used-car sales, to name a few. But does such a need exist for ethics in science? If so, what are the tenets of its ethics code, and have there been violations of that code? The immediate answer to the first question follows simply from recognition that science is a human endeavor. If science were the strictly logical pursuit of knowledge isolated from the rest of society, then its rational character and lack of societal spin-off would obviate a concern for issues of ethics. We argued in Chapter 2, however, that science is not a purely rational and logical activity. Inspiration, creativity, and intuition are essential in making scientific progress. In addition, the data that are used at the forefront of science can be murky and open to multiple interpretations, as well as be subject to poorly controlled, or even unknown sources of error. The emotions of the scientist, plus other interests that are related to commercialization, tenure and promotion, or simply the ego of the involved scientists, can open the possibility of unethical behavior. In short, again, science is a human activity, and its practitioners are human. Beyond this, science necessarily is strongly intertwined with society; civilization today is shaped by scientific innovations and their technological upshots. The fruits of science have commercial and military applications that might or might not be in line with the motives for which the science was initially pursued. For all these reasons it is important to consider the ethical facet of our behavior as scientists.

The principles that guide ethics in research depart in no substantial way from those that guide other aspects of life. Rather, they are tailorings of those more general ones. Such guidance amounts to sets of *dos* and *don'ts*. The former, which emphasize the positive, are pointed toward steps that aid the success of not only the individual but also of colleagues in the profession and of the profession in general. Unfortunately, the stronger more negative guidance, the *don'ts* – which amount to a set of rules – tend to receive more emphasis largely because they can be stated more explicitly. Perhaps another reason for their greater emphasis is the protection of the practitioner: in any profession and in life, one's reputation is something that can be given away only once. It is more productive to think of ethics as a set of virtues, rather than a set of rules that must be obeyed; the value of living these virtues is captured in the following words of Emerson (1841):

> In a virtuous action I properly am; in a virtuous action I add to the world.

Principles and standards for ethics in research can be divided into those that govern the style in which research is carried out (Sections 8.1–8.6) and those concerned with the content of research (Section 8.7). For both, the main rules for carrying out research are the same as those in other aspects of our life: *respect the property of others, be honest, share appropriately with others, acknowledge others' contributions fully, treat others as you would like to be treated, and take pains to consider the consequences of your actions.* These topics are treated in the following sections. Much of the material in this chapter is inspired by the report "On being a scientist; responsible conduct in research,"[1] published by the National Academy of Science. That report gives a clear and succinct summary of the ethical aspects of being a scientist.

In some areas of research, the safety of health of living beings can be compromised. An example is experiments with animals that could be harmful to them. Clinical trials with patients could directly damage the health of patients, or they might entail the withholding of proven treatments from patients by exposing them to unproven ones instead. Sometimes experiments involve risks of exposure to chemicals or radioactivity. Even though precautions are taken to minimize the risks and suffering for test animals, patients, and researchers, these risks often cannot be eliminated altogether, and a difficult balance must be struck with the benefits of the research. Because of the special character of such dilemmas, we don't discuss these further in this general treatment of ethics. We refer the reader to texts on this topic listed in Appendix A.

8.1 RESPECT THE PROPERTY OF OTHERS

In carrying out research, it is crucial to respect the intellectual property of others. There are many ways, some of them subtle, in which one might take unfair advantage of such intellectual property. It is not possible to give a complete overview of all the ways in which

[1] http://www.nap.edu/html/obas.

one can misuse the results of others, but the following examples might give you an impression of the proper spirit in which to carry out research – a spirit that, throughout your career, will enhance your sense of satisfaction and merit the respect and appreciation of your community.

It is perfectly acceptable to use information from others as long as you give them proper credit. When you quote someone else's text, it is essential to make it clear that you are actually or closely quoting their work, and it is appropriate to make explicit reference to the original text. When using a result or idea proposed by somebody else, refer to that person's contribution. Ideally, one can refer to a journal article, book, or conference abstract. When this is not possible, you can always make reference by using the term *personal communication*.

Giving proper credit for the scientific contributions of others is much more than a matter of courtesy. The citations in our papers can have tangible consequences for the authors cited. Because of the ready online availability of the number of times that a paper is cited (Section 9.1), it is convenient for review panels and university administrators to use the number of citations as a measure of the influence of a researcher's contributions. A well-cited author is better known in the scientific community and thus has a better chance of receiving funding for future research. The impact of research, as measured by citations, can be a major factor in decisions involving promotion and tenure. Properly citing the work of others thus goes further than solely giving due credit; it can influence decisions that have consequences for their career.

At times you might learn of somebody's idea before it is published in the literature. This could happen through some conversation or when reviewing a research proposal or article. It might well be that, because of this insight, you see a productive line of research for yourself. You then have two options. The first is to refrain from carrying out that research or postpone your work until the original idea is published. The second (preferable) option is to discuss your plans with the person who gave you the idea and perhaps start a collaboration.

Reviewing proposals is a sensitive issue because, in their proposals, fellow researchers describe research that they intend to do but have not yet tackled. In the "Instructions for Proposal Review," the National Science Foundation therefore states:[2]

> **Your obligation to keep proposals confidential**
> *The Foundation receives proposals in confidence and protects the confidentiality of their contents. For this reason, you must not copy, quote from, or otherwise disclose to anyone, including your graduate students or post-doctoral research associates, any material from any proposal you are asked to review. Unauthorized disclosure of confidential information could subject you to administrative sanctions. If you believe a colleague can make a substantial contribution to this review, please obtain permission from the NSF program officer before disclosing either the contents of the proposal or the name of any applicant or principal investigator. When you have completed your review, please be certain to destroy the proposal.*

Such requirements apply in practice as well to reviewing articles for scientific journals. Surprisingly, these guidelines do not state that you cannot use the information and ideas you obtain in the proposal for your own research. That, however, should be understood implicitly; it holds without saying. The simple rule and guiding principle "respect the property of others" captures this ethical rule much more succinctly than could any legal jargon.

8.2 BE HONEST

Apart from respecting the intellectual property of others, it is essential to be honest in research. In research, data are crucial for assessing whether or not a theory agrees with reality. In Section 7.7 we noted the work of Popper (1965), who states that a confrontation of theory with reality, through taking measurements, is needed to test if a theory can

---

[2] http://www.nsf.gov/pubs/1997/iin121/od9708a.htm.

be falsified. His point is that in the absence of our ability to establish absolute truth, this is the best we can do as scientists. The use of data makes sense only when we can trust those data. Of necessity, we often must rely on the results of experiments carried out by others, and it is essential that we can trust that those experiments and results have been truthfully reported.

That point is also made by Bronowski (1956), who states in his wonderful book *Science and Human Values* that science is based on two pillars: trust and dissent. We consider the issue of dissent in Section 8.4. In an ideal world we would carry out and report our science in such a way that a colleague can repeat the experiment that we have done. In practice, this often is extremely difficult. It might not be realistic to repeat an experiment because of the large economic costs and logistic complexity, a measurement could involve an extremely rare event, the number of degrees of freedom in carrying out the experiment might be so large that exact replication of the experiment is unrealistic, or it could be unethical to repeat an experiment unnecessarily, for example in patient trials. The repeatability of measurements therefore often is an illusion. In the absence of the ability to redo an experiment, we have to trust that colleagues truthfully report their work.

Following are various mechanisms, some of which are not purposeful, by which scientific results can be reported in a way that departs from reality.

- *Unintentional errors.* Making errors is unavoidable. Most often these errors are caught in time and don't influence the work presented to others at scientific meetings or in the scientific literature. It does happen, occasionally, that undiscovered erroneous results are presented. If done in good faith, this poses no ethical violation, but it is incumbent to correct this error publicly whenever possible, and as soon as possible.
- *Negligence.* Some researchers are sloppy, to the point of being negligent. Laboratory procedures, such as calibrations, might not be

carried out properly. Data could have been improperly archived, or worse, might have been unsuitably screened before being archived. When science is not carried out with proper care, its results can easily be in error. A slippery slope exists when data are being screened or selected before they are included in the analysis or shown to colleagues.

- *Fraud.* Sometimes the pressure or the temptation to obtain certain results is so large that results are fabricated by changing measurements, by inventing measurements rather than actually taking them, or by leaving contrary measurements unreported. Such scenarios are not at all far-fetched. The popular press and scientific literature show that this type of scientific fraud regrettably occurs on a regular basis, and the known cases likely form the tip of the iceberg. This can be motivated by commercial interests where the economic incentives prevail over scientific integrity (Washburn, 2006).

- *Self-delusion.* Not all falsification, improper selection, and interpretation are done intentionally. Often, the pressure or the personal desire for a particular outcome of an experiment can bring about to a self-delusion that leads to an unfaithful reporting of the work performed or an improper interpretation of experiments. It can be unclear in specific cases whether this type of self-delusion is intentional or not.

In reality, it can be difficult to categorize a given instance of erroneous reporting of results into any particular one of these four categories, but this is not needed: the point here is not to be judgmental. Rather, it is to provide insight into the various factors that can lead to incorrect reporting of results. In the following we explore, further, motives that can prompt forms of unethical behavior in reporting of results.

When carrying out research, one often has a desire for experiments to yield specific results, either to support a preferred theory or, conversely, to help refute an unwanted one. This desire that the data work out in a specific way can influence, even subconsciously (e.g., through exaggeration), the way in which we deal with data. Moreover, regrettably it has happened that researchers have consciously falsified or purposely shaded data in order to arrive at the desired results. In the worst of cases, some researchers have simply invented the data without even carrying out an experiment. (Nothing subconscious about this.) No matter how infrequent, such acts, when exposed, provide argument for those who wish to discredit one side or the other of politically charged topics of scientific relevance, such as evolution theory and climate change, or to discredit the entire scientific endeavor itself.

Falsifying data or inventing data are not the only ways in which data can be used unethically. Data can be abused in more subtle ways. By using measurements selectively, one can nudge the data to support, or disprove, a specific theory. In any experiment, however, it can happen that certain measurements are influenced by observational errors that could render them useless for interpretation. Such erroneous measurements should legitimately be discarded. Thus there exists a gray area in which it is not objectively clear whether or not a measurement ought to be used. The general rule is to discard data only when you know and can defend why these data are invalid for your study or analysis; that is, discard data only when you are able to identify the source of error in the data. Even then, it is appropriate to inform the reader about the type of data you have discarded and about your rationale for having deleted them.

In one prominent example from the 1980s, a research group claimed that when a chemical was diluted in water to such an extent that, statistically speaking, no molecules remained in the dilution, this extremely dilute solution still produced a biological effect from the molecules that originally had been dissolved in the water. The idea was that the chemical had left an "imprint" on the water in the dilution. The prestigious journal *Nature* published this claim despite the editorial board's skepticism about it. A team under the direction of the Editor-in-Chief of *Nature* subsequently investigated the experiments. The team concluded and reported that the researchers had selectively discarded "bad measurements" (Maddox *et al.*, 1988). The researchers had retained only those measurements that supported their theory, which by this action became a self-fulfilling prophecy. As seen in this not-so-gray example, the temptation to discard certain measurements can be great, especially when much is at stake.

Be aware of this temptation, sometimes subconscious, to nudge observations until they give a desired outcome, i.e., one that you have the "deep-down feeling" to expect. Avoid being influenced by a subjective desire for specific results, and attempt to remain as objective as possible, a challenge at times given the complexity of many experiments and numerous degrees of freedom for analyzing data.

Injecting bias into experiments can be avoided by carrying out "blind experiments." In such experiments, measurements or samples are randomized in a way unknown to the researcher so that this person cannot know what is the expected outcome of each measurement. This approach is routinely used in patient trials in which a new drug is tested on patients. In its simplest form, the group of patients is randomly divided into one subgroup that receives an experimental drug, and another that receives conventional treatment. It is, of course, documented which patients receive the experimental drug, but this information is disclosed to neither the patient, the researchers, nor the caregivers of the patient. In this way, bias in the care given to the patient, the expectations of the patient, and the interpretation of the results of the treatment is avoided. Such an approach is not limited to experiments with human subjects, but can be applied to a wide range of experiments. Many types of scientific investigations, however, are not amenable to being conducted in such a blind manner, and often require subtlety of approach and ingenuity to minimize contamination by bias.

8.3 STAND UP FOR YOUR SCIENTIFIC INTEGRITY

Back at the level of conscious choices, economic and political pressures offer frequent temptation to shade interpretations, for scientists and engineers both in industry and academia (Laughlin, 2002). A prime example arises when an individual is invited to serve as an expert witness in a litigation that requires scientific or technological expertise. The expert witness is seldom hired by the court. Rather, one side or the other in a legal case pays the often large salaries of the witness – and has the expectation that the witness will give scientific judgement that is favorable to the client. Examples include expert-witness psychologists who offer their informed opinion as to the mental state of an individual on trial in a capital case, and expert-witness engineers called on to offer their informed opinion on the safety of a proposed site for construction of a hydro-electric dam or a nuclear power plant. Almost invariably, that informed opinion just happens to favor the interests of the client who pays the salary of the witness.

Here is an example that has indeed taken place and is quite subtle, but involves litigation whose outcome can mean hundreds of millions of dollars to one party or another. Suppose that two major oil companies are partners in the development of an offshore oil field. The percentage of revenues allotted to each company is ascertained on the basis of the percentages of the subsurface reservoir that lie beneath the areas whose production rights are owned by these two companies. Expert witnesses with recognized geoscientific credentials, whether they are academics or consultants from industry, are asked for their opinion, on the basis of the geologic and geophysical data, as to what are the percentages of the hydrocarbon resource beneath the acreages owned by the two companies. Perhaps only one or two percentage points difference are in contention. This seems like not much, but for a 200-million-barrel field,[3] with oil prices at $90 per barrel,[4] 2% of 18 billion dollars is $360M, sufficient to offer handsome salaries to the differing expert witnesses, even after the lawyers have been paid. But with what degree of certainty and to what level of detail can a geoscientist state that a given percentage of the resource lies beneath the property of one party or another? In such a situation it can be tempting to let economic interests prevail over scientific integrity.

Next is an example of a dilemma that might be faced by a young scientist or engineer with an environmental consultant company that

---

[3] 200 million barrels of oil seems a lot, but it covers the consumption in the USA for only 10 days.

[4] With the recent fluctuations in the price of oil it is difficult to keep this number up to date.

evaluates the safety of a given site against actual or potential environmental damage. Suppose that the client company has a large interest in a study stating that the site is safe, and the young employee is pressured by a superior in the company to slant his findings in favor of the wishes of the client. Perhaps in this instance the shading is relatively minor. At this early stage in the young employee's career, capitulation to an opinion against his judgment can start him down a slippery slope in future studies he does for his employer, but how is he to go against the wishes of his superior? Our guidance is to urge that the young employee stand up against the pressure and either give the honest opinion that goes contrary to the client's wishes or resign from the company. Those stances are a tall order for a young, new employee, but ones that he will not later regret and that might even serve as inspiration for others in his company.

To these examples involving economic consequences, we might add those of political pressure exerted toward slanting a scientific finding in one direction or another, or to suppress scientific findings in order to suit some political purpose, say those related to the extent that contributions to climate change are man-made. Whistle-blowers in government laboratories must have thick skin to endure the accusations of those who might be involved in suppressing such findings. Again, the choice to stand up in favor of the scientific evidence as you find it can be a most difficult one, but capitulation to pressure to go against or suppress your scientific findings can send you down a path that you might well regret in the long term.

8.4 DISAGREE RESPECTFULLY

The friction of debate creates illumination.

Goswani, 1995*

As mentioned above, Bronowski (1956) makes the case that science is based on both trust and dissent. We have discussed in

---

Section 8.2 the importance of trust; here we expand on the value Bronowski attaches to dissent and its implications for our behavior as scientists.

Science would not move forward much if we all were always in agreement because then we would be happily reinforcing the same ideas, a situation not conducive to growth and innovation. Growth and innovation are spurred by individuals who are dissatisfied with current explanations or who disagree with the current state of knowledge; they feel impelled to find a new theory or interpretation. Kuhn (1962) points out that this process of changing the paradigm can be long and difficult.[5] It is inherent in the process that scientists at times find themselves in disagreement about their work, and it is such disagreement that finally resolves itself in a new interpretation that pushes science forward.

Much can be at stake in science. The reputation of scientists can hinge on their being "right." Promotion and tenure decisions usually depend on the quality and level of innovation of one's science, and research funding is often allocated on the basis of the reputation of scientists. Also, patents and related commercial spin-off can depend on the work conducted. It is in this tense and competitive playing field that different opinions about scientific issues need to be resolved. Under these conditions, exhibiting grace and respect while in opposition can be difficult, but it is doable and always worth doing.

The point made by Bronowski (1956) applies to democracy as well, since this political system is also based on trust and dissent. Trust is needed for there to be constructive dialogue and negotiation. Dissent is needed in a democracy to avoid the complacence that would lead to the perpetuation of the status quo at the price of failing to seek societal innovations or to respond intelligently to changing external conditions. It is thus for the same reasons that science and democracy

---

[5] A case in point is the theory of continental drift as proposed by Alfred Wegener in 1912. This theory found little acceptance until the theory of plate tectonics was introduced in the 1960s. It took more than 50 years and some impressive new data sets for this paradigm shift to take place.

are based on trust and dissent. Let us not dwell on the slippery moral slope of politics, and instead return focus to the role of dissent in science. As scientists we have to balance a solid conviction of our views on a scientific issue with a mind open to changing our views when needed. Striking this balance is not easy. Some people are too easily swayed by the views of others, while others stick to their guns even in the face of the most convincing evidence.

Swaying too far toward a stubborn conviction of one's views can lead to unnecessarily defensive behavior, hostile public attacks, unfair reviewing of papers or proposals in order to block progress of competing groups, and even back-stabbing of colleagues. These are but a few examples of ways in which a scientific dialogue can degrade into unethical behavior. Fortunately, such behavior is not the norm; most scientists are fair and reasonable and can work out scientific disagreement in a constructive way that pushes science forward. Given the openness typical of the scientific community, the small fraction of scientists who have difficulty disagreeing with grace generally pay a high price for this behavior. Disagreeing with respect is not only valuable for dealing with colleagues, it is essential for creating the constructive dialogue needed to push science forward.

8.5 AUTHORSHIP ISSUES

I was surprised when Professor Long named me as a co-author of his gravity map project. Sharing the spotlight with an undergraduate was not a common thing to do, and the generosity of this gesture left a lasting impression. It helped me understand that giving people ownership and credit in a project is the surest way to guarantee its success.

Rutt Bridges, personal communication

Different people have varying styles in their approach to research. Some are individualists who like to do things on their own whereas others have a stronger group instinct and prefer to share their work with colleagues. When working with others, it is proper and right to share the fruits of your work with those who helped you plant and grow them. Authorship of scientific articles is nevertheless a recurrent source of conflicts and frustrations among research colleagues.

Objectively, the rules of the game are simple. Those who have contributed significantly to scientific work should be co-authors, while those who have contributed less so ought not be among the co-authors, but should have their contributions acknowledged. Thus, for example, a fellow student who gave you a great idea one night over a beer should be among the co-authors when this idea is crucial for the work you present. This rule also implies that senior researchers should not muscle their name in among the list of authors when they have not contributed sufficiently, and they certainly should not let their hierarchical power prevail over their scientific contribution by insisting on being first author when someone else played a much more essential role in the research.

This brings us to the often touchy point of who should be credited as being first author of a publication. The individual designated as first author plays a central role because in practice the outside world will perceive that person as the one who contributed most to the work presented. We typically use the following simple rule for making the choice: the first author is the person who has put in the most time and effort. Usually, but not always, this is also the individual who actually writes the paper. Since writing the paper is a significant time investment, this rule usually resolves any conflicting interests among those who co-author a scientific paper.

Unfortunately, things are not always so clear-cut as presented above. There is no objective standard for deciding who has contributed significantly to the research that leads to a publication. It might happen that someone had made a casual comment that was important to you but did not constitute a major scientific contribution or entail time investment toward the research. Do you ask that person to be a co-author? There is no clear rule for deciding this. Fortunately, there are ways of giving your colleagues the credit that they deserve other than making them co-authors; you can refer to their publications, refer to them by using the phrase *personal communication*, or use the acknowledgments section to give them the credit they merit.

Because no strict laws or rules exist for deciding on co-authors, you will need to let your common sense prevail. Our view is that it can be distinctly beneficial to be open to liberally sharing the authorship of articles with others. *When in doubt, share.* First and foremost, this is a friendly and courteous gesture to your colleagues. Second, this fosters a trusting relationship with your colleagues that in turn can be essential for future collaborations. Third, when you share your work with others, they will be more inclined to share theirs with you.

Regarding this third point it is helpful to make a distinction between your short-term and long-term goals. In the short run, it might seem better to be possessive of your work and to portray yourself to the rest of the world as the single parent of a scientific discovery. In the long run, however, it is an advantageous strategy to share the results of your work with those who helped you carry out that research. We urge that, in general, you let your long-term interests prevail over your short-term ones.

When one of us (Roel Snieder) worked for his Ph.D. thesis at Utrecht University in the Netherlands, his advisor, Guust Nolet, always refused to be co-author of any article with his students. Professor Nolet would say he felt that he had not contributed sufficiently to be co-author, but this usually was not true since he played an essential role in giving his students' work the direction it needed. This generous attitude came from a desire that his students should have the visibility they needed in the scientific community to embark on a successful career.[6] This is an attitude that we think is beyond the call of duty of a research supervisor. It, however, serves to illustrate another reason for not being too possessive about authorships. As a researcher

---

[6] When Guust Nolet read an early version of this manuscript, he commented that the pressure to publish that pervades the present American funding system has made him less inclined to waive the possibility to be co-author on papers of his students. Also, in Europe the funding system has become much more competitive since the 1980s.

who supervises students or junior researchers, you are a role model for those for whom you are responsible. When your juniors respect you, they will watch you closely; your behavior often is an important guiding principle for them. In your daily behavior, you reinforce the positive or negative habits of your juniors. Therefore, your actions need to reflect your mental attitude about the things that are important in a healthy research environment. Sharing the fruits of the work is one these things.

8.6 INTERACTING WITH OTHER PLAYERS

As we noted in Section 2.6, science has many different players, with overlapping emphasis. Universities focus on both pure and applied research, government laboratories aim at a wide spectrum of both pure and applied research as well as on the application of knowledge, and industry has activities that range from mostly applied research to the application and commercialization of knowledge. These different types of organizations can interact fruitfully because their activities partly overlap, as shown in Fig. 2.4, although their missions differ. If the activities of these different organizations did not overlap, there would be no basis for interaction. It is this overlap that creates opportunities for collaboration and additional value. The differing missions of these different types of organizations, however, can lead to tensions and ethical dilemmas. We describe some of these in this section.

These different players should be aware of their specific mission and restrict their activities to those that fall within their mission. For example, broadly speaking, universities are there for teaching and carrying out innovative research. Universities should not compete with industry when it comes to the application or commercialization of knowledge because these activities are beyond the mission of universities. This is all the more so for universities that are publicly funded. With use of public funds, these universities could find themselves in a position of undercutting prices that industry would have been able to charge for certain activities, an abuse of public funds.

For similar reasons, government laboratories should not be involved in the commercialization of knowledge beyond helping to foster such commercialization in the marketplace.

Industry can be an important sponsor of valuable academic research because of the funding it can provide, the data and facilities that industry can make available, and the interaction with talented and enthusiastic researchers within the industrial organization. If precautions are not taken, however, this collaboration can potentially lead to conflicts of interest.

Common sources of tension include restrictions that might be imposed by the industry partner on the publication of research. In the academic community it is essential for both the career of the faculty member and successful completion of graduate studies to publish the results of research, and to do so on a timely basis. Publishing might or might not be part of the corporate culture of an industrial laboratory; in any case it is not generally of prime importance. Open publication might even go against the commercial interests of the industrial sponsor because it can make the discoveries available to competing companies. It is crucial that the different partners in joint academic/industrial research projects discuss this issue up front. Specific guidelines as to what is or what is not an acceptable compromise are neither easily nor generally defined. In our own industry-sponsored research, we have accepted the imposition of delays in the publication of research that are of the order of just a few months, but refused longer delays or further-reaching restrictions on publication rights. It is especially important that graduate students be free to include material used in their research and results of sponsored research in their M.Sc. or Ph.D. theses since restrictions to do so could jeopardize their ability to graduate in a timely manner.

In industry-sponsored research, patents and intellectual property can pose dilemmas for university partners involved in joint research. The perceived economic value of intellectual property could lead to restrictions on the academic partner's ability to publish results of research, especially in the thesis. From the ethical standpoint, the advice "share appropriately" is likely to be the best course of action. In our contracts with industry we have used the simple recipe that the findings of the research are the intellectual property of the persons who made the discovery. This could mean that the findings are the joint intellectual property of both parties.

A patent is a formal claim regarding intellectual property. Patents pose an interesting ethical dilemma because they exclude others from using the results of the research without consent of the owner of the patent. This practice is contrary to the desire in academia to make the results of research available openly. Because of the increasingly legal and defensive attitude of society, it is increasingly common to protect intellectual property with patents. Although the restrictions imposed by a patent might appear to be a negative consequence, the filing of a patent can actually be useful in making the results of research available to society. It can take a large investment in time and money to push the results of scientific research to the stage where it is available to benefit society. Turning the prototype of a technological invention into a product that can be marketed can take a significant amount of engineering, testing, and certification. In biomedical applications, for example, the required trials needed for certification of a new drug can be extremely time-consuming and costly. Companies, and their investors, in practice, are unwilling or unable to invest in the commercialization of research results without reasonable assurance that they are the sole owner of the intellectual property. This shows that the application for patents leads to an unavoidable tension between the wish to make research results widely available to the community and the value of providing the exclusive rights that often push the research toward a commercial product that could benefit society.

It can happen that industry-sponsored research yields results that are unfavorable for the industrial sponsor. Patient trials, for example, might show that a new drug is not superior to an existing one, or research might show that a new theory simply does not work. In such situations there can be pressure from the industrial sponsor to prevent publication of the research and its results. Worse yet, there even could be pressure to manipulate the data in such a way that they portray a more favorable outcome for the sponsor. Maintaining scientific integrity can be a challenge for the scientists involved, especially when the pressure comes from a large sponsor of research. As scientists, we should hold that scientific integrity must prevail over commercial interests. Washburn (2006) gives a distressing account of the impact of increasing commercialization of research and teaching at universities.

As sketched above, the collaboration between academia and industry poses many challenges, in particular when facing ethical decisions, in addition to the opportunities that it offers. It is prudent to discuss these issues openly at an early stage so that they can be resolved lest they become acute in a later stage of the research. An almost unavoidable complication exists because discussion about the terms of the research contract usually requires the involvement of lawyers. Because of their legal expertise, it is necessary that they be included in the discussion. Keep in mind, though, that the lawyers work for their client, and are paid to hammer out the best possible deal for their client. Scientists should be careful not to sign away their rights too easily, and stay attuned to the potential ethical consequences of restrictions that might be imposed by the contract.

If your intention is a career in academia, with the above warnings in mind you might become wary of industry-sponsored research or collaboration with industry. Keep in mind, though, that industrial laboratories employ world-class scientists with great expertise in their field and who can be wonderful collaborators. Moreover, joint research with industry offers the availability of unique facilities and data that can be a great boost to research. As seen in Fig. 2.4, the range of activities and expertise of the different generators and users of knowledge overlap, suggesting great benefit in collaboration. For their mutual benefit as well as for the benefit of the scientific endeavor in general, groups with differing missions and goals must find a way to collaborate and do so in ways that do not infringe on their individual missions and goals while upholding high standards of integrity.

8.7 ETHICS OF THE CONTENT OF RESEARCH

Creativity unguided is a two-edged sword. It can be used to enhance the ego at the expense of civilization. One must apply creativity with wisdom.

Goswani, 1995

The ethics of science concerns not only the way in which we carry out science; it also governs the content of our scientific effort. In science, we aim to increase our understanding of the world. This includes the development of methods that place increasingly more of the world in our hands. The development of a drug can help to eliminate or suppress a certain disease; exploration of the Earth's resources raises our standard of living; development of electronics makes information more widely available; and the ability to control the fission of atoms creates the possibility to generate electrical power and to make weapons of mass-destruction. In general, science and our scientific activity increase the power that mankind has over the world.

This power can be used for good or ill. Sometimes the distinction between good and bad applications of science appears clear and simple; more often, however, the distinction is blurred. For example, while nuclear physics has given us the ability to generate massive amounts of climate-friendly electricity, it has also introduced unprecedented means for mass destruction in the world. Such distinction between good and bad applications is not always as clear-cut as it might appear. Nuclear-generated electric power is considered a bad development in the eyes of many because problems with the disposal of radioactive waste can leave our children with a heritage of highly toxic waste products. Conversely, during the Second World War the development of nuclear weapons was, for many of the researchers involved, considered an honorable activity because it appeared to be the only way to prevent the nightmare scenario that the Nazis would dominate the world by being the only power in possession of nuclear weapons (Rhodes, 1995). Physicist and molecular biologist, Leo Szilard, who was among the early leaders in the development of the atomic bomb, while looking back on his stance stated that *"We turned the switch, saw the flashes, watched for ten minutes, then switched everything off and went home. That night we knew the world was headed for sorrow."*

More subtly, hydro-electric power, which previously seemed to most people to be a common-sense approach that puts readily available water resources to use for the benefit of mankind, is now seen as having introduced hidden costs. These costs (many of which are not yet fully comprehended) include alteration of the ecological system, environment, and climate, as well as forced relocation of large numbers of people and encouragement of over-production in agriculture and of uncontrolled over-development in housing.

Thus it is not simple to ascertain whether a given technological development is good or bad for mankind. This judgment is personal and can depend on the context in which it is made. To make matters worse, more than in the past it is often difficult to foresee what will be the consequence of a scientific development and its technological offspring; "unintended consequences" are a common by-product of research in our technological world. These unintended consequences can be either positive or negative from an ethical point of view. Given the large potential for good or harm in technology and science, it is incumbent to put care into creatively envisioning, as best as possible, plausible consequences.

Science itself will not tell us what are the ethical values of its results. Jonas (1982) argues that all science can be used in a negative way, and that scientists have no control over the way in which their work is used. He uses this to argue that further scientific developments are undesirable. We feel that this conclusion is overly negative, since his point of view underestimates the benefits of science for society. Nevertheless, regardless of the complexity of foreseeing the consequences for our work, it is crucial that each of us develops and acts on an "inner standard" of what is acceptable and what is not. It is imperative that we develop a *noble purpose* for our work. This noble purpose is that aspect of our work that makes our daily activities rise beyond the level of "just doing a job." The noble purpose is what provides *meaning* in our work, as discussed in Section 6.5.

Such a noble purpose is appropriate for more than scientific work. It is applicable to most of our activities. For scientists working in academia, for example, teaching is an integral part of their activities. A noble purpose for education was formulated as follows by Myers (2001):

> I think we have the responsibility to insist that education is more than learning job skills, that it is also the bedrock of a democracy. I think we must be very careful that in the race to become wealthier, more prestigious, and to be ranked Number One, we don't lose sight of the real purpose of education, which is to make people free - to give them the grounding they need to think for themselves and participate as intelligent members of a free society. Obsolete or naïve? I surely hope not.

Whether or not you agree with this noble purpose of education, it is essential for the betterment of the world that we, as scientists, who indeed are among society's leaders, develop our own noble purpose for our work, and that we use this as a guiding principle in our activities. The formulation of a noble purpose is neither obsolete nor naïve; it forms an integral part of what gives essential meaning to our work and to our lives.

9 Using the scientific literature

I must say that I find television very educational. The minute somebody turns it on, I go to the library and read a book.

Groucho Marx

Traditionally the academic library has been the repository for the archiving of books and journals for scientific research, also offering a place for reading and study. It is much more than that today. By making available large and readily searchable databases of books and publications, these libraries have come to offer researchers the capability to search the scientific literature with remarkable efficiency, retrieving relevant publications electronically when possible.

Indeed, retrieval of scientific information is increasingly being driven by electronic tools and information technology. While this development opens up new possibilities for the efficient search and retrieval of information, it does so in the face of a new problem: the amount of information available is vastly larger than what the individual human mind can process. It therefore is essential to access this superabundance of information in ways that actively supports the research. The options available nowadays could be bewildering, making it important to be aware of, and use, the right tools for gaining access to the appropriate information. Most academic libraries offer valuable assistance and suggestions through their websites. Moreover, typically the staff of these libraries have the expertise as well as the desire to offer advice and share their expertise with those seeking help.

Whatever the purpose or stage of research – embarking on a new line of research, addressing specific research questions, or scanning the breadth of fields that are closely aligned or remote from your primary focus of research – a literature search is essential. This is discussed in Section 9.1. New scientific developments become available continuously, mostly through the scientific literature. Section 9.2 offers advice on how to keep up with this stream of information. In Section 9.3 we discuss ways to use a personal database containing bibliographic material.

Traditionally, scientific information was disseminated via books, journals, scientific conferences, and personal contacts. With the advent of the internet, a new source for the retrieval of information has become available. Search engines, such as *Google*, are extremely powerful for searching the internet using specific keywords or phrases. Nevertheless, you are probably aware, or should be, that anybody can post anything on the internet; the medium has no quality control whatsoever. The information posted by research groups and individuals on the internet is usually not peer reviewed and can be unreliable. Just the same, search engines such as Google, with their capability for finding pointers to scientific information, can be extremely helpful for getting started on, or pursuing, a new aspect of research.

9.1 DOING A LITERATURE SEARCH

Before embarking on a research project, it is of fundamental importance to find out what research has already been done in the chosen field of research. Duplicating the work of others is a waste of time unless it is done intentionally with the goal of reproducing or challenging the work of others. In order to contribute to the state of knowledge in a field, one needs to know what is the current state of knowledge, what are its limitations, which methods have been used to move the research forward, and which groups are pursuing similar research. To be adequately prepared for original research, it is therefore necessary to carry out a literature search.

As mentioned above, a modern literature search is not carried out by randomly browsing books and journals, although such an activity can lead to interesting findings. Instead, much of the sought-after information is accessed electronically through websites and manuscripts in electronic form (often in *pdf* format). Despite the electronic character of information retrieval, we still use the traditional jargon of doing a *literature search*. Searching through the literature can be done either backward or forward in time, the two approaches offering complementary value.

Searching backward in time is most familiar and easiest. Suppose you have an article about a topic that interests you. Then, the references cited at the end of the article point to work deemed relevant and carried out prior to the article's publication. This allows backtracking through previous research in order to follow the reasoning, successes, and stumbling blocks that others had encountered. The paper trail (well, perhaps not paper these days) that starts in the reference list often leads to related papers that can offer valuable insight. Papers disseminated in electronic form sometimes contain links to cited papers, making it convenient to retrieve these papers and to archive them electronically.

Searching forward in time reverses the familiar search direction. Often, one would like to know what research has been carried out since the publication of a paper of interest. One way to search forward in time is to look for articles by the same author published at a later date. This method is not sufficient, however, because other authors might well have subsequently published relevant articles on the same topic. Searching forward in time can be done effectively with the *Web of Science*, a database that provides lists of articles in which specific publications are cited. This database provides an efficient way to find publications that have followed up on the research described in the primary article. The database lists the names of authors who have cited every paper produced by a given scientist over a specific time frame.[1] The Web of Science, however, is expensive to purchase, so small universities unfortunately might be unable to afford access to this database.[2] Also, be aware that, although the Web of Science lists

---

[1] University administrators and funding agencies sometimes use (or abuse) this information as a measure of the quality of researchers. This is one reason that it is important that your work be such as to merit future citations by others.

[2] The fact that information is accessed in electronic form does not mean that it is free. In fact, for-profit publishers are effective in extracting a price for the retrieval of the information they provide.

citations in a broad range of journals, it does not cover all journals; hence the information it provides is not necessarily complete.

These search methods use the scientific literature as a string that connects research over time, a powerful means for tracking down relevant work. With modern information technology, however, it is also possible to search the scientific literature directly. Dedicated bibliographic databases are available for finding and making available publications of a certain author, or publications that contain specified keywords in the title or abstract. These databases typically are devoted to specific scientific disciplines and offer an excellent tool for finding material within those disciplines. They, however, generally are available only through subscription. Universities usually have subscriptions to the bibliographic databases that are most relevant for the institution. These databases are typically accessed through the internet, although sometimes it is necessary to go to an academic library to gain that access. Most of the databases have the option to send the results of a search to your computer by email, a particularly useful feature for retrieving a large number of references since they are then directly available in electronic form. As discussed in Section 9.3, this is convenient for both creating a personal archive of the information and efficiently generating reference lists when writing papers.

Many databases can be searched through the internet. While general search engines can be used, they are not geared towards retrieving scientific publications and thus the search needs to be supplemented by more-specific ones. Google has a specific search engine (*Google Scholar*)[3] dedicated to retrieving scientific publications. Moreover, numerous databases exist for searching the scientific literature in specific fields (Table 9.1 gives examples). A subscription

---

[3] Websites are likely to change rapidly with time. For this reason we refrain from giving web addresses. It might therefore be necessary to search for the web address of tools, databases, and services listed here. Also, new websites, databases, and information-management tools are continuously being developed, making it worthwhile to search for these tools every so often.

Table 9.1. *Examples of databases for searching the scientific literature*

Field	Database
General	Science Direct
Biology	Biological Abstracts
Chemistry	Scifinder Scholar
Computer Science	Computer Database, Inspec
Engineering	Compendex
Geoscience	Georef
Mathematics	MathSciNet, Locus
Medical	Pubmed, Medline
Physics	Web of Science
Ph.D. theses (general)	Dissertation Abstracts

is typically needed to search these databases. Academic libraries usually carry such subscriptions, and these databases can be searched through the library website.

Most Ph.D. theses contain an overview of the current state of research in a field, along with an extensive literature list. Because Ph.D. theses are typically written in a less concise form than are most research papers – providing more background and technical detail – theses can often be more accessible for the non-specialist than are research papers. For this reason a recent Ph.D. thesis can form a good starting point for research, especially for researchers new to a field. The database *Dissertation Abstracts* can be used to search effectively for Ph.D. theses on a particular topic. With so many reference databases, it can be confusing as to which ones to use in a search. The reference librarians in academic libraries are happy to offer help and advice on how to proceed.

In addition to databases dedicated to finding, publications are other sources of information. Professional organizations publish journals and have effective tools for searching their publications. Also, many individual researchers and research groups have websites that give an overview of their research, and often it is possible to retrieve their publications in electronic form from those websites. This is most useful when you know which are the leading groups in a certain field of research. If you are uncertain as to where to search, then your adviser or colleagues likely can provide pointers to websites worth visiting. Be aware, though, that material posted by individuals or research groups might not have been peer reviewed, so that their content can lack the quality control that most journal publications provide.

Both search engines for the internet and databases for scientific literature search for words or phrases. The databases usually contain the following information about a certain scientific contribution: author, journal, volume number and page numbers, title, keywords, and text of the abstract. In order for you to find an article, the relevant words must be in the title, keywords, or abstract. Thus for your publications to be found electronically by others, it is important that well-chosen words that characterize the content of the paper be featured in those locations. Simply choosing the right words can greatly contribute to the visibility of your paper within the scientific community. If keywords have not been supplied by the author, the journal of the database publisher might assign them. Since you are likely to make the best choice of keywords for your paper, you are well advised to provide them whenever the opportunity exists.

The information posted on the internet, the search tools, and the databases sketched above are wonderful resources for retrieving information. Much of that information is relatively recent, which is good for helping you be aware of what research has recently been done and what is happening now. The short shelf-life of this information, however, can easily give the impression that the only relevant activities are those taking place right now, or in recent years. We want to stress here the value of reading classic papers in a given field. The original papers are sometimes written in a style and with jargon that is difficult to read these days, but often the classics in a field are written with superb clarity and reveal the line of thinking that has led to a scientific breakthrough. As described in Chapter 2, the path of science is often tortuous. One can gain great insight from learning about the origins of a particular scientific finding or development. No amount of internet searches or impressive-looking websites can convey the insights – and admiration for pioneer scientists – that you gain from reading the original descriptions of breakthrough research.

9.2 KEEPING UP WITH THE LITERATURE

Likely, you are not the only person on the planet who works on a specific research topic. Awareness of the work of others can save much time and perhaps embarrassment. Others will have disseminated the results of their work in scientific papers and at conferences. To keep up with current research, it is crucial to follow the scientific literature of the field in which you carry out research.

Moreover, it can be worthwhile to read the scientific literature in areas outside your specific area of research. Not only can such reading be a source of inspiration, but work carried out in other fields might have a bearing on your own investigation. Another important reason for having awareness and understanding of research outside your own specialized field is that, when you complete graduate school and apply for a job, potential employers often want to hire you to work in a field of research other than your specialization. Your chances of getting a job offer will be enhanced if you have at least an idea of what goes on in neighboring fields. Beyond that, the more flexible you are in the research you can do, the greater are your options for an exciting and intellectually challenging career.

Although this advice is logical and straightforward, it is not so simple to implement. So much is being published that you can spend all your time reading the work of others and fail to do original work at all. Reading new issues of even a single journal can absorb an inordinate amount of time. Reading all the journals in your specific field is a yet more daunting task, so when scientific papers from other fields are included in your mix, you might develop the feeling that you are drinking from a fire hose. Clearly, a strategy is needed to make this flow of information manageable. The following advice might be helpful.

- Decide which articles in a journal issue you want to read by going through the following steps. Read just the title first, and decide whether you want to spend more time on that paper. If the answer is yes, read the abstract and decide again. A subsequent step could be to study just the figures and captions. If you want to go into further detail, you can still read the paper in different ways. You can read it "diagonally," not worrying about parts that you don't understand, or you can grind your way through the paper until you fully understand it. In practice, you will probably choose to compromise between these two approaches. In this way you can scan or read papers at a progressively increasing level of detail, while making a conscious choice about how much time to invest in a given paper.
- Make a choice as to which journals you wish to follow and at what level of detail, and which ones you will ignore altogether. We recommend following at least a few journals in your area of specialization, plus general scientific journals, such as *Science* and *Nature*. The general journals can help you keep abreast of breakthroughs in related fields of research that might interest you, while the specialized ones will, you hope, keep you up to date in your own field of research.
- It can be useful to follow the work of specific people or research groups that are leaders in your field of interest. Often, this information can easily be accessed through websites that show recent research, published papers (and often unpublished ones as well), and perhaps data or software that can be helpful. Visiting such websites on a regular basis can be highly effective for following recent developments in your field of research.
- Many databases in science and technology have an alert service that sends email messages to the subscriber based on keywords chosen by that user. This tool makes it possible to, with little effort, be aware of publications in certain research fields spanning a broad range of

journals. It is important to choose the alert criteria with care; otherwise alerts add unnecessary clutter to the glut of information that comes your way.

Even when taking these steps, it is impossible to read all articles that are relevant for your research. When working in a research group, it is possible to assist each other by making colleagues aware of publications that might be relevant for them. Another approach to sharing information is to organize a *journal club* wherein members in turn present overviews of articles to the others in the club. Doing this on a regular basis helps alleviate the load of following the literature, and assists in broadening the scope of those in the research group. The discussions that ensue often are fun, give insight and inspiration, and can be a starting point for collaboration.

9.3 MAKING A DATABASE OF REFERENCES

Reading many papers and books is a integral part of doing research. Many of these sources will later be used in the reference lists of publications. When reading papers, you might imagine that you will later remember them, but over the time-scale of years, or even months, one tends to forget details of papers read. We, of course, seldom remember page numbers and volume numbers. You will be surprised that you at times also forget names of authors and the journal in which a specific article was published. It is important to be able to access this bibliographic information rapidly in order to retrieve information from a specific author or research topic. For this reason, it is convenient to create a database of references. Traditionally, index cards were used to store bibliographic information, but nowadays an electronic database offers many advantages since it can be rapidly searched in many different ways (e.g., by author name, title, keywords, journal type, and year of publication).

With a suitable database, you can include these references in your manuscript by using a few simple commands, with no need to tediously enter details such as journal volume and page numbers for each new paper that you write. This habit is called *cite as you write*; it saves time spent on re-entering bibliographic information in reference lists of papers, and, perhaps more important, it avoids the errors inevitably made when re-entering such information. So, the choice to be made when deciding whether to use an electronic database of references is: do you want to enter bibliographic information once, or do you want to redo this every time you cite a paper in a publication?

Much software is available for archiving bibliographic information. It is important to use software that can easily be connected to the word-processing tools used. Many scientific papers are written with Microsoft Word (or equivalent free programs such as Open Office), while scientists in the physical sciences often use LaTeX. Since Microsoft Word (typically abbreviated as *Word*) and LaTeX are the most common word-processing tools, we focus on these programs.

LaTeX relies on a database of references stored in *BibTeX* format. The structure of this database is illustrated in Appendix C. Every entry in the database has a field – the key – that gives the name of that publication. For example, suppose a reference has been given the key `Jones2001` in the database. The citation to this reference in a manuscript can be made by including the command `\cite{Jones2001}` in the LaTeX file. The citation to the Jones (2001) paper is then automatically included at the place in the text where the cite-command is entered, and the bibliographic information drawn from the database is included in the reference list. This makes it possible to cite while you write, a habit that leads to huge savings in time. Different journals and publishers use different formats for the reference lists; you can even specify which format to use for the reference list that is printed in each particular paper.

The Word program has a similar feature for efficiently including references in a manuscript, provided that a database and database program are used that can interface with Word. One such program is *EndNote*. This program can read a database stored in *Refer* format, which is described briefly in Appendix C. The choice of specific database format is not critical. Presently, numerous available free software programs can convert one type of database into another, and many database-management programs have the ability to import and export a bibliographic database in a variety of different formats.

Regardless of which word-processing package you use, and in which format the database is stored, it is essential to use keywords to help find your way in the database. Well-chosen keywords are essential for retrieving the information in the database efficiently and for reducing the chance of failing to include important publications.

As described in Appendix C, by using any text editor one can create a database of publications in BibTeX or Refer format in the form of a simple text file. This, however, is not the most effective way for creating and using such a database; software specifically dedicated to this task is available. A screenshot of the database program *Jabref* is shown in Fig. 9.1. This particular program is designed to archive

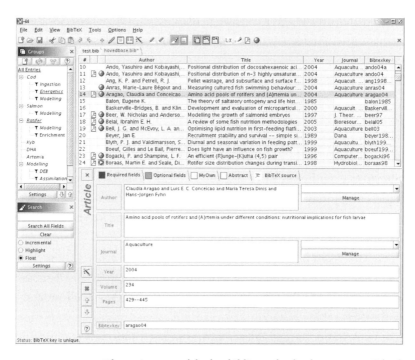

FIG. 9.1. The main menu of the free bibliography database program Jabref.

and search bibliographic databases in BibTeX format. We illustrate the advantages of a dedicated database program with this screenshot. All references are visible in the form of a table in the top-right. The large window in the bottom-right (labeled "Article"), which shows the bibliographic information for a selected publication, can be used to enter or change this information. This simple way of modifying information obviates the need for editing BibTeX files by hand. Given the somewhat complex format of BibTeX files (see Appendix C) and the unforgiving nature of LaTeX for format errors, this tool can help avoid time-consuming correction of such errors. It is also possible to store the abstract of a publication and to add notes, which can serve as a reminder of the contents and significance of a publication. As shown by the window labeled "Groups" on the top-left, publications can be grouped, making it possible to organize the database by topic. The "Search" window on the bottom-left allows searching of the database using some combination of the date, author name, and keywords. To the left of the list of authors in the window on the top-right are icons indicating those publications that are available as pdf files. Thus, this database program can archive publications in pdf format, making it possible to retrieve these pdf files easily.

The program Jabref, as shown in Fig. 9.1, is one of numerous choices available. The commercial word-processing software, End-Note, is popular with users of Word. The free software *Zotero* is conveniently embedded in the web-browser *Firefox*, which offers the advantage of being an easy interface between the bibliographic database and information retrieved through the internet. Nowadays, many different database programs are available to choose from. Here are some considerations to keep in mind when choosing a database program.

- *Format(s) used.* The program Jabref, shown in Fig. 9.1, is dedicated to archiving information in BibTeX. This is handy for users of LaTeX, but not for those using Word. Most programs, including Jabref, can import and export a database in a variety of different formats; it makes sense nevertheless to use a database program that is designed to handle the database format that is actually used.

- *The ability to interface with the word-processing tool used.* A database program is not useful if the program, or the database that it creates, cannot interface with the chosen word-processing package. For LaTeX users this is not an issue because LaTeX, and the free companion program BibTeX, can retrieve bibliographic information from any file in BibTeX format. For users of Word, it could be necessary to use a database program that can actually interface with Word.

- *Price.* Some database programs, such as the popular program EndNote, are developed for commercial purposes and can be quite expensive. Currently, many alternatives, such as Jabref, *Pybliographer*, and Zotero, are freely available.[4] The savings in time and increased effectiveness for using and retrieving bibliographic information might make the expenditure for commercial software worthwhile, but, given the numerous free programs available, you might want to explore the value of using one of them.

- *Platform dependence.* Computers that scientists work on each require a specific operating system. Most common are *Windows*, *Unix*, and *Linux*. Software runs only on operating systems for which it is designed. This is not an issue for programs written in *Java*, which operate under any operating system. Other programming languages, however, are not so flexible. Be sure to choose software that actually works under the operating system that you use.

- *Reliability.* A database of references is valuable for research because it archives information collected over a long time. Searches that are not foolproof can lead to serious mistakes and omissions, and files that are contaminated with formatting errors can require much time for correction. For these reasons, any database program must have been well tested and reliable before you give it the important task of

---

[4] These are just examples of programs; many others exist. Given the rapid pace of software development it is worthwhile to search carefully before choosing a program.

managing your bibliographic information. Be especially wary of beta versions of software, newly developed software that has not yet been fully tested by a large users' group.

In general, it is convenient and cost-effective to use the same software as that employed by other members of your research group. Using the same tool makes it possible to share information, to help each other in solving problems, and to make most effective use of the tool. Choosing a common format for storing the database makes it easy to share information, thus increasing the amount of information available to all members. Sharing bibliographic data is even more convenient when using internet-based citation-management software, especially when collaborating with researchers at other locations. It will be helpful for you to discuss the use of database software with your adviser before choosing a specific database program and spending significant time on setting up your bibliographic database.

A bibliographic database captures a wealth of knowledge and information painstakingly accumulated over many years. Losing that information would be a serious setback in research; at the least, recovery could involve a major time investment. As with all software and data, it is essential to have adequate backups of the bibliographic database. This small investment in time and hardware can help avoid major problems and setbacks.

10 Communication

Our task is to communicate experience and ideas to others.
Niels Bohr

The most basic answer to the question "why communicate?" is simply that scientists are human beings, and communication is one of the defining characteristics of being human. More specifically, here we elaborate on the value, both to individual scientists and to their profession, of communicating the findings of research.

Likely, a major factor that initially drew you into and has kept you in your chosen scientific career is the simple joy of intellectual discovery, of uncovering new ideas about how the physical world works. For many highly capable scientists, the internal satisfaction they experience in such discovery is reward enough. It is time-consuming and can be laborious, indeed burdensome, to detour from the scientific pursuit to share one's findings with others, through either publication in peer-review journals or presentation at conferences. Painful as it might be to devote energy to that detour, for several reasons expect communication of results to be of great value to you both personally and professionally. More broadly, publication is the breath of life for future progress in the sciences.

As emphasized in the next section, communication is difficult, and *effective* communication, whether in writing or delivered orally, is especially difficult. There must be *some* reward for taking such pains to share what you have been doing. Potential personal rewards are many. The obvious ones are recognition and visibility of your scientific contribution, and the associated prospects for research funding and employment. Recognition and respect gained from those in your scientific community can be highly rewarding. Professional societies annually award honors to those who have made significant contributions to their field, and receiving such recognition can be most satisfying. That is likely not to happen, however, if you simply plug away in your laboratory or at your computer, taking quiet delight solely in the discovery process itself. Job opportunities often arise from sharing research with colleagues and potential employers, and the chances for attracting research funding also increase when one's work is well known within the scientific community.

Whatever one's personal inclinations toward devoting the energy necessary to sharing results of research, the care given to presenting the product of research will prove valuable both personally and professionally. Success in turning a tenure-track position into a tenured one, for those who have chosen an academic career, is often dependent on the highly pressured requirement of "publish or perish." The counterpart situation in industry is the weight that management puts on employees' ability to communicate with others within and outside the company. Whether you are in academia or industry, several less pressure-induced benefits also accrue to you personally when you communicate your ideas and results openly with others. When you share, others return the favor in kind, providing feedback on your work as well as offering their ideas and wealth of experience, both of which can be helpful to the next stage of your research. Especially during early stages of your career, the respect that you earn from having effectively communicated your work can serve as a magnet that draws others to follow your work closely, with the prospect that fruitful collaborations develop.

Whatever anxiety you might experience in preparing a manuscript or preparing and presenting a lecture, the elevated sense of satisfaction that comes from having conveyed your ideas to others, especially when they are conveyed with clarity, can be highly rewarding.

Perhaps surprisingly, the labor you put into careful preparation of your findings for presentation to others can be of large benefit to the quality of your research itself. More often than not, the effort of thinking carefully through how to present your ideas to others gives clarity to your own understanding of the work you've done. Your effort to explain your work to others can help crystalize and sharpen your thinking, often yielding new ideas that inspire new direction to your research.

Those are benefits that accrue to the individual scientist from his or her communications with others. More broadly, science can advance only when its practitioners share their research with others. For centuries, from before the time when the search to understand the natural world was given the name *science*, publication has been the primary source for archiving advances in knowledge. Today, publication of papers in peer-reviewed journals and of abstracts for presentations are the means for adding to that invaluable archive. Science stands on the shoulders of more than giants; its advance depends on the cumulative understanding of lesser mortals as well, and that understanding is preserved in the vast literature available to researchers. The combined shoulders are broad indeed.

10.2 COMMUNICATION IS DIFFICULT

Individuals, scientists and non-scientists alike, span the full spectrum from those who love the process of expressing their thoughts and ideas in writing to those for whom writing is close to the most dreaded of activities. Whether one actually enjoys writing and giving oral presentations or whether these are excruciatingly painful undertakings to be put off in favor of doing just about anything else, communication is difficult. Again, *effective* communication, in particular, is especially difficult. We are not speaking here of casual communication, such as in email or telephone conversation with a friend, although likely we've all experienced occasions in which misunderstandings have developed from failures to be clear and precise in our intended message.[1] Here, however, we focus on the tasks of writing technical papers and preparing technical presentations.

---

[1] Communication between management and employees in organizations of all types suffers notoriously from failure at being effective. In our experience, when an organization has achieved the reputation for having just moderately acceptable internal communication, someone in management has been doing communication back-flips to accomplish even that.

The first difficulty for those who do not look forward to writing is in just getting started. Procrastination is a favored practice among such people. Whether or not you're in this category, the advice that we offer is the same as that which some of the most noted authors follow – just sit down and write. You might first jot down key thoughts, then try your hand at a broad outline, and then add detail to it. But, now what words do you choose? The helpful step here is to simply start putting words on paper. Don't worry about how well they are chosen. Just blast ahead, applying words to paper or word processor. Later, you can revise (and revise and revise). You get started, however, by putting words out there so that you'll have something to revise.

For all writers and presenters, whether or not the task seemingly comes easily, no matter what the form of communication – one-on-one discussion, tackling agenda items in a business meeting, or presenting scientific findings to an audience – effective communication requires that the sender and the receiver(s) are attuned to one another. We can offer no more valuable advice about your writing than that you *know your audience*. For your communication to be effective, you must put yourself in the shoes of the audience; that is, *be* the audience. Because this is not as easy as it might sound, we recommend that you at least work thoughtfully at *trying* to put yourself in their shoes. Also, given the difficulty of doing this with success, don't worry excessively about the need to be in tune with your audience during your early process of blasting words onto paper. As you revise, however, have the audience, and your connection with them, paramount in your mind. With attention to this crucially important perspective throughout each revision, you can bring your paper progressively in closer touch with the audience.

After all, isn't your purpose, whether in writing or presenting, that your ideas get through to your audience? Not only must you present your material so that the audience understands what you are saying, you should take care to make what you have to say *interesting* to them. How else can you hope to capture and keep their attention? Readers and audience members typically have many other things to do and that can occupy their mind, making it all too easy for their thoughts to drift away.

The following advice again holds, whether your paper is to be written for publication or presented orally. Be on the lookout for aspects of your material that will be a challenge for the audience to understand, and pitch the communication at a level that you thoughtfully perceive to be right for that audience. A complication is that the audience often is not homogeneous, so the ability to handle the complexity of information received can be expected to vary among its members. This puts you, the speaker or writer, in the difficult position of having to find a tone and level for delivery of the message in a way that fits a diverse group of people. As we will emphasize again in the section on oral presentation, a message comes across most clearly when it is aimed at a level that is slightly *lower* than what we might think the audience can deal with. The simple reason for this is that, because of the time constraint, the audience for oral presentations must devote more intense and immediate concentration than does the reader of a paper. For reasons that will be discussed below, however, it can also be helpful to target even written papers somewhat below the level you might at first think appropriate.

In the following sections, we will particularize this overarching advice – that you consider the audience foremost – to both papers for publication and those to be presented orally.

10.3 WRITTEN COMMUNICATION

Repeating from earlier, for many people (the present authors included) writing is a daunting, difficult, and even onerous and painful task. This section aims to help overcome some of the difficulties that impede the work of writing and enhance clarity in the writing.

10.3.1 Prepare before starting to write

Before writing a scientific paper, you of course must know what are the scientific findings and insights that are to be communicated and have clarity about the research methods, the results, and their meaning and interpretation, as well as an understanding of how these results fit into the larger picture in the field of research. These issues pertain to the content of the presented work, but more is needed. You must also have clarity about the style and level of the presentation.

Again, know the specific audience for your work. A scientific article is written for fellow specialists who know the jargon of the field, whereas a popular text aimed at high-school students (and at scientists with expertise outside your specific discipline) should be devoid of technical jargon and unnecessary abbreviations. For whichever type of paper, your communication will benefit from having an easygoing style, avoiding the temptation to impress with elevated verbiage. Aim for clarity, foremost.

Before writing, it is helpful first to consider the following questions:

- Why do I want to write this article or report?
- For whom am I writing?
- How long will the manuscript be?

The first question helps set a goal for your writing effort, without which your actions might easily lose direction. The second one helps to define the tone and style to use in your writing. The third question addresses possible length restrictions imposed by, for example, the target journal. Moreover, in considering your audience you might well choose to impose your own length restrictions so as to increase the effectiveness in getting your message across.

Once you have answered these questions, you are still not yet ready to start writing. Broadly, two different aspects of writing are best kept distinct and in proper order. First, you must decide what is the message; only then do you have the message that needs to be translated into words. Reversing the order of these two steps or intertwining them is a common source of difficulty for many authors. Although finding the right words might never be easy, you might be surprised to find how very blocked you can be in finding the words for formulating a message on which you have not yet settled. The key to breaking out of this deadlock is to put the *what* in front of the *how* of the writing. *First decide in detail what you want to write; only then start writing.*

Writing is made easier by taking the following steps. First, write a rough outline of the manuscript. This initial outline contains the main structure of the manuscript and the main points that you want to get across in the different parts of the paper.[2] Second, refine the outline by inserting more detail, and continue this process of making the outline progressively more detailed until you get to a point at which it is completely clear to you what you want to say, the order in which you will make the different points, and the examples and illustrations you want to use. For a technical publication, this means that, before tackling the writing, you have written out all equations explicitly and that you have prepared all figures, tables, and diagrams. For a popular presentation, this means that you have prepared all the illustrations and that you know how you intend to get the message across to your target audience.

For papers with co-authors, at this pre-writing stage discuss ideas about the outline with your fellow authors. We advise that graduate students bring the adviser into the loop as well, even if she isn't one of the authors. Consider the following nightmare-scenario. A graduate student painstakingly writes a paper or a chapter for a thesis. She gives this lengthy manuscript to the adviser, who, because of its length and perhaps weakness of organization, takes six weeks to read it. This is a recipe for then having to do a drastic revision of the structure of the manuscript, possibly requiring that most of it needs to be rewritten. Such an inefficient use of time can be avoided by discussing the outline of the paper *before* the actual writing has begun. It, in fact, is generally easier to discuss the content of a paper on the basis of an outline than by working with the written manuscript.

---

[2] Often, the primary stumbling block in getting started arises from difficulty in setting out an organization of the paper that flows. The outline is the tool for avoiding this potential source of confusion.

The discussion of a paper that has already been written can easily be cluttered by arguments that involve style and grammar. Also, an outline is simply much more quickly and easily understood than is a lengthy manuscript, thus offering the opportunity for more rapid feedback from co-authors and adviser.

Our research often follows a complicated path. It is common to discover during a research project that the approach that was taken is not the most direct way to solve the problem. As stated in Chapter 2, this is an unavoidable result of the unpredictable nature of original research. It is tempting to describe in a paper all activities that were carried out in the course of the research. For the reader, however, learning about the tortuous path taken during the project is typically not useful. In fact, the research results are usually conveyed in the clearest way by following the shortest and most logical path of reasoning. To have to leave out the description of all research activities that had been carried can feel frustrating, especially for junior graduate students. Remember, though, that a research publication is not a work report; its contents serve the purpose of conveying scientific results and insights rather than accounting for all the work that has been carried out. Your readers will appreciate your having gone to the trouble to help take them efficiently through your work and message.

> The secret to being a bore is to tell everything you know.
>
> Voltaire (1694–1778)

10.3.2 Writing is an iterative process

Writing is most effective when it is started after completion of the preliminary work described in the previous section. If your outline is sufficiently detailed at this point, you know quite well *what* you want to say; hence you can focus on *how* you will say it. If you discover at this point that you are still not clear about what you want to say, you likely need to go back to your outline and make it more detailed. Note that in this process we once again are going through the steps *thoughts* $\Rightarrow$ *words* $\Rightarrow$ *actions*. When writing scientific papers, you might find it easiest to write the sections covering research methods and results first, before writing the introduction and conclusions, and finally the abstract. These three all-important parts of the paper depend on the core of the paper – the text describing research methods and results.

As with so many others skills, practice is crucial for good writing and for becoming comfortable with facing the writing task. Therefore, rather than shunning the task of writing, use it as an opportunity to practice a skill that you will value having cultivated. Moreover, actively seek suggestions and feedback from individuals who you know to be good writers.

We believe that there is no book about writing style in English[3] more valuable than the classic book, *The Elements of Style* by Strunk and White (2000). Fortunately, this helpful and much-valued little book, first written in 1935, is not likely to go out of print, having been enhanced by E. B. White, a noted novelist.[4]

Much more than a book of grammar, it is a book on writing *style*. Among different ways of writing that all follow rules for proper usage of grammar, punctuation, and words, authors still have much latitude in the ways in which to express their ideas. Perhaps most novel and valuable about the Strunk and White book is its emphasis on style that helps with clarity and readability of the written work, the primary goal being writing that *holds the interest of the reader*. As a reader, you've all too often experienced having to wade through cumbersome, tedious text, but you might not have realized that the reason for the difficulty has been failure of the author to convey the message directly,

---

[3] For other languages, proper choice of writing style plays a similarly valuable role in aiding the effectiveness of writing. Quite possibly a counterpart to the concise and friendly *The Elements of Style* can be found in those languages.

[4] The popularity of this book among noted authors has been so large that the book recently has inspired a new version, *Illustrated Elements of Style*, with witty illustrations interspersed to amplify the message of the original book. This gem of a book has even inspired an *opera* of the same title.

concisely, and clearly. If you take to heart the suggestions in this little book (and it is little), you will develop a resourcefulness and economy in your writing – an ability to convey *more* with *fewer* words.[5]

Much has been written specifically about technical writing (see suggestions for further reading in Appendix A). Our bias is that good technical writing is none other than good *writing*. Therefore, if you learn to write with economy and clarity, your technical writing will be the beneficiary. One tip that we have taken to heart is to write in the *active voice*. This helps the reader of a passage know what was done and who did it.

So, with all these aids to effective writing, composing the paper should be easy, right? Unfortunately, the task can still remain daunting, as holds for any creative endeavor. The following two-part recommendation might be of help in overcoming "writer's block." You've seen the first part earlier in this chapter. With your refined outline in hand, *start writing*; "blast away." Don't worry about the rules for good writing style you've absorbed from the Strunk and White book. Don't even pause much over misspellings of words. Rather, write fast, almost in a stream of consciousness (but based on the organization you've worked out in your outline). The words you are pouring onto paper or into the word processor constitute just your first draft. Be conversational, as if you are sharing your ideas with a friend. (Friends don't worry much over the style in which you form your sentences, and don't even know how you've spelled your spoken words.) The second part of this recommendation is the stage most critically important for producing quality writing – *revise, revise, revise*.

In revising, one does much more than cleaning up typos, misspellings, and grammatical blunders, and more than putting

---

[5] Perhaps most advantageous to developing strong and engaging writing is that you have an "ear" for the language, something that is more difficult (but not impossible) for those for whom English is not their native language. We can suggest nothing better for anyone wishing to develop that "ear" for a language than to read respected modern literature in that language.

into practice the elements of clarity, economy, and emphasis that constitute effective writing style. Your writing will benefit from doing something that, although it might seem simple, is not easy – *putting yourself into the shoes of the reader*. We mentioned this previously, but are restating it because of its fundamental importance to all that you write. Your writing is done not for the sake of you, the author. It is to convey, as effectively as possible, a message to your readers. By *effectively* we mean again with clarity, economy, and emphasis, all of these traits focused on holding the interest of the someone who is *not* you; that someone is the reader.

Likely, you have spent many months in developing your ideas, testing them, and obtaining your results, often following blind alleys along the way and coming to your ideas via circuitous routes. You are intimately familiar with the subject of your paper. Expect, therefore, that, no matter how capable are the intended readers of your paper, they lack your familiarity with the subject. Conveying your message to someone who is not you, requires, above all else, an *empathy* with the reader. As you attempt to identify in your mind who is that reader, avoid defining your perceived class of readers too narrowly. The more broad your intended readership, the more likely you will minimize use of jargon and abbreviations (which are inherently exclusionary) and the more care you will likely take in laying out helpful background information.

Having an empathy for the reader means caring for and giving attention to the reader — to the extent that you view the reader as a friend. This friend has not traveled the (likely winding) path that you've taken in your research. Now you would like to lead your friend down a path to the results of your investigations, and you would like to do so in such a way that she can readily follow her own way there, retracing steps if necessary. Through your empathy, you will mark the junctures in the path with clarity, and show her the shortest path that preserves that clarity, avoiding the lengthy, perhaps confused path that you had taken in your research. Moreover, you will naturally wish to share some of the excitement in your research efforts and findings; to do that, you will wish to maintain her interest throughout the duration of the trip along the path. With such an outlook as author, your research and writing can be fully appreciated, understood, and valued by your scientific community.

Once you have written and revised (and revised) the manuscript, your work is not yet done. Any writing that we do always leaves room for improvement. Among other things, we understandably are often blind to our own shortcomings, especially those in work in which we have invested much energy and perhaps have developed a vested interest. Moreover, even with the most conscious effort to put ourselves in the shoes of the reader, we are not the reader. After you have written and worked over a manuscript, you have several ways in which to get the distance from the work necessary for you to improve it further:

- First and foremost, show the manuscript to one or more caring colleagues and ask for feedback. Because the friend who critiques your paper is not you, she offers the perspective of a reader who lacks your familiarity with the details of your research. She is representative of your readers, so you can expect that points in your paper that are unclear to her need clarification for the readership at large. One recognized value of the extensive peer-review process followed by respected scientific journals is that your paper benefits from the perspective, insights, and questions of others. Consider your invitation for a friend to critique your paper as the first step in the peer-review process. The more thorough her critique, the better is your friend.
- Put the manuscript away for a while. Time can give the distance needed to see many of your own mistakes or gaps in your argument. This is one reason why writing abstracts, articles, or proposals just before the final deadline is generally not a good idea.
- Some authors find it useful to read the manuscript aloud. Hearing the manuscript in spoken words gives a different perspective from that of seeing the words on paper. This new perspective can make it easier to identify and make improvements.

You might be pleasantly surprised by how much your perspective on the text you've written changes when you follow these three steps. This minimal investment of effort can help enormously in improving a manuscript. Use the opportunity.

As a related aside, scientific journals have good reason for inviting respected individuals to review submitted papers. Aside from a reviewer's important feedback on the scientific merits of a paper are any indications in the review that she has misunderstood part of the author's message or that she is put off by lack of clarity in the writing. Authors sometimes become defensive when receiving criticism about any aspect of the manuscript, whether it relates to content or presentation. Understandable as is that reaction, we encourage you to consider any criticism by a reviewer as a welcome aid to improving your manuscript. Suppose that a reviewer disagrees with some content of your paper. Instead of taking such disagreement as an indication of a failing on the part of the reviewer's intelligence, ask yourself (1) "have I failed to convey my ideas with sufficient clarity?" and (2) "do the reviewer's concerns reveal something that I hadn't thought of?" In either case, view the critique as that of a friend (even if you have reason to believe otherwise) who is a representative member of your potential readership. With that perspective, you can then use the critique to the benefit of your final published paper.

With your nearby colleagues as the first line of reviewers, their feedback can benefit your paper in yet another, highly valuable way. If your manuscript is improved as a result of their feedback, the journal's reviewers will find your paper more readable and, yes, enjoyable, thus increasing the likelihood that your manuscript will receive a favorable review.

10.4 ORAL PRESENTATIONS

Findings included the following: impact [of lectures] appeared to be greatest during the first 5-minute portion of the presentation, with impact sufficient to cause students to [recall] about 35% of all ideas presented; impact declined, but was relatively constant for the next two 5-minute portions, and dropped to the lowest level during the 15- to 20-minute interval; ... presentations with more than about 40 ideas or bits of information were likely to be less efficient, with impact dropping off as the information load increases.

Burns, 1985

Oral presentations at conferences and in seminars, which are a key method for timely communication of scientific developments to colleagues, constitute an essential part of the scientific endeavor. Expect, and we encourage you to seek, opportunities to share your research findings in oral presentations to members of your scientific community. Yet, for many scientists, presenting research papers orally at such meetings is stressful.[6,7] Our primary advice for those struck by some degree of trepidation over speaking in front of large and even modest-size audiences is that you recognize that members of the audience are your *friends*. They are there out of interest in hearing what you have to *say*, not from a desire to judge you. Think of your presentation as a conversation with perhaps an individual friend with whom you would like to share the excitement of the work that you have been doing.[8]

For various reasons other than trepidation, the quality of oral presentations at meetings varies greatly, with remarkably large room for improvement. The oral presentation is not only a channel for communicating, indeed advertising, scientific results, it can also leave a powerful personal impression – for good or ill – on the audience. Some speakers present their work in such a sparkling way that they come

---

[6] While we hope that our advice can help reduce the large stress that some presenters experience, there is value in being *keyed up* when giving a talk. This state helps the presenter maintain thoughtful concentration that can be lost when she is totally casual in giving the presentation.

[7] The website http://www.healthpsychology.net/Helpful_Skills.htm offers helpful skills on basic relaxation and thought-challenging techniques often used by psychologists for relieving stress.

[8] In conversation with friends, we often don't speak in full sentences. Likewise, while you should plan thoughtfully what you want to say in a presentation and how you want to say it, you should not feel the need to construct complete sentences. Periodic simple phrases, rather than complete sentences, in a talk can give emphasis that helps maintain audience interest.

across as creative professionals, while others shoot themselves in the foot by their effort to impress their audience with incomprehensible scientific jargon, or through lack of care to take simple steps that can help the audience understand as much of the message as possible, and as clearly as possible.

In oral presentation, just as for written communication, it is essential to match the level and content of the presentation to the audience. The large challenges specific to oral presentations are (1) the limited time allotted to the speaker for getting her point across and understood, combined with (2) the typically limited attention span of audience members. The quote by Burns, above, focuses on the problem of limited attention span to be expected of students attending a 50-minute course lecture rather than that of a conference audience hearing a 15- to 20-minute lecture. The issue of capturing and retaining an audience's attention remains as well for the typically shorter conference lecture.

Paraphrasing a noted exploration geophysicist, Carl Savit, "Your oral presentation to a large audience will have been effective if 20% of the audience got 20% of your message." Although given tongue-in-cheek and perhaps with some exaggeration, Savit's comment gives an idea of what the audience is up against in attempting to gain maximum information content from oral presentations, and the daunting task confronting the presenter wishing to convey her message effectively.

Even more so than with written communication, a good starting guideline is to imagine the audience's level of understanding of the material, and then aim slightly lower. Your audience will be grateful that you reached out to them in this way. There's no need to fear that the audience will think you simple-minded by aiming slightly below their technical level.[9] People don't enjoy it when they have to stand at the tips of their toes for an extended period of time; they truly appreciate an effort by the speaker to explain the work clearly.

---

[9] Judgment, derived from experience, is required here. An audience can be offended if the technical level at which you pitch your story is significantly beneath that of the typical audience member's understanding.

We have found that even specialists in the audience, who can readily handle a presentation at the highest level, usually are appreciative of a presentation that is well constructed to help the majority in the audience understand the message.

10.4.1 Preparing an oral presentation

Some speakers give the impression that they improvise freely and are completely at ease when speaking in public. For some people this activity indeed comes naturally, but for most of us it takes much work to put together an effective presentation.[10] This especially holds for scientific conferences, where oral presentations must often be of short duration. It is not uncommon at scientific meetings that the time between the start of successive presentations is as short as 15 minutes. With the time required to change speakers and for discussion, this leaves only about 12 minutes of speaking time for each presentation.

Preparing an oral presentation therefore requires care in planning. Not unlike that of preparing a written paper, the form and content of the presentation will follow from first addressing the questions *who will I be talking to?* and *what is the message that I want to get across?* Of importance to the success of the presentation is the rehearsal process. Among several advantages of rehearsals are that they help the speaker be more comfortable with the presentation, they make it possible to get the timing of the presentation right, and they offer an opportunity for feedback from others. After several rehearsals, the flow of the presentation becomes more natural, creating a sense of certainty for the speaker about exactly how to give the talk.

Timing of the presentation is important. It is considered bad form for a speaker to go over time, so at scientific conferences the chairman of a session will cut off a talk that takes more than the allotted time. Because rehearsals help ensure that the timing will be just right, a speaker who takes too much time either has done this

---

[10] Even for those who have the reputation for smoothly lucid, apparently extemporaneous delivery, you might be surprised by how intensely they prepare their talks.

on purpose or has not taken the trouble to prepare and rehearse the presentation. In less formal settings, such as seminars, the audience might interrupt the speaker often by asking questions. If this might make it impossible for the speaker to deliver the intended message, it is appropriate to say something to the effect, "I am happy to take questions for clarification, but I would like to defer any discussion to the end of my presentation."

The rehearsal is most efficient and helpful when it is done with an audience of colleagues, some of whom have worked closely with the speaker, and others who are less familiar with the research work. The former group can be especially on the lookout for key points about the research that the speaker might have failed to cover or convey effectively, and the latter have a distance from the presentation that makes them more representative of the audience expected in the forthcoming conference. Any confusion that they have over points made in the rehearsal is indicative of confusion that the conference audience can be expected to encounter. Your rehearsal audience can also provide important feedback on the style of presentation – are the slides readable, did you speak too fast, did you have any strange habits such a playing with your keys, did you maintain the interest of the audience? Practicing with your peers also helps you be more comfortable in delivering the presentation to others.

10.4.2 The use of projection

Most scientific presentations are nowadays supported by graphics, and LCD projection has almost universally replaced the use of slides and overhead transparencies. Processed multi-dimensional data (e.g., medical images or seismic sections), graphs that portray an idea or measurements (e.g., bacterial population trends), and photographs of data subjects (e.g., photo-microscopy images) simply cannot be portrayed effectively through words; graphics slides that show such material are thus essential to a talk. Slides that contain just text can be useful as well, primarily in helping an audience remember key points. A disadvantage, however, of the overuse of text slides – particularly those that are excessively wordy – is that the projected material tends to distract the listener from you, the speaker. Most important in your talk are your message and the effectiveness of your delivery. Therefore, except when you need to show graphic or pictorial material, you want the audience to have its attention fixed on you, your facial expressions, and body language.

Stated differently, slides should be an aid to the speaker, not the other way around. Care must be taken that the projection is a tool that supports the presentation rather than a source of distraction from the speaker and the content of the message. Regrettably, speakers have grown so used to giving presentations structured around LCD projection that these days few are able to give a presentation without the use of projected material. For a topic that lends itself well to the following exercise – and some do – try giving an oral presentation using no projection at all. You might learn that, for other presentations, you can use far less projected material than you might previously have thought.

Graphic material nevertheless is an essential tool in oral presentations. Of crucial importance, the material shown must be clear and readable, with special attention given to audience members in the back of the lecture room. We recommend the *1/20 rule*, which states that the lower-case fonts should not be smaller than 1/20th the height of the slide window projected on the screen.[11] When preparing slides or figures, this font size, when viewed on paper, seems huge, but it will work well for your presentation.

The readability of text varies for different font choices. For some fonts, such as *Times Roman* (the font used for the text in this book), the thickness of lines that make up the letters depends on the line orientation. (Take a careful look at the letters in the text here.) *Helvetica* and *Arial* are fonts with lines of constant thickness independent of

---

[11] Note that we are not recommending a particular numerical font size; font size in a slide scales with size of the projected image, which one hopes is sufficiently large for the room. Also, do not trust that a font size that looks sufficiently large when you preview the figure on paper, or even on your computer screen, will be acceptable when projected in a lecture room. In general it will not be.

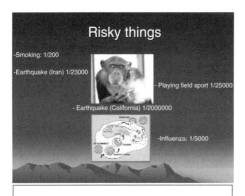

FIG. 10.1. Two examples of slides that aim to convey the same information.

line orientation. In general, and particularly when the projection is less than ideal, fonts with a constant thickness are more readable than are those with variable line width, and boldface versions of them are more readable yet in slides.[12] Figure 10.1 shows an example of two slides that attempt to convey the same information. The fonts in the top slide are too small for the projection used in most seminar rooms, and the font used (Times Roman) has a variable width, degrading readability of the letters. In contrast, the bottom slide shows a font size and font type (Arial) that can be read comfortably, even when sitting in the back of a lecture hall.[13]

The slides in Fig. 10.1 have a width of 6 cm on the page. Viewing these slides from a distance of 60 cm gives the same view angle

[12] Likewise, curves with relatively thick line width are much more readable and comfortable to the eye than are narrower ones. Lines that look sufficiently wide on paper are generally too narrow for comfortable viewing in projection. We suggest viewing different line widths from the distances mentioned in the following paragraph.

[13] Also, as an aid to the audience, it is generally best to maintain consistent font size throughout a slide, and from one slide to another.

as that for slides projected on a screen 2-m wide and seen from a distance of 20 m (the back of a classroom) or as that when projected on a screen 6-m wide viewed from a distance of 60 m (an auditorium). In all cases, the width of the projected image is about 10% of the distance to the screen. Any lettering and graphic contained in Fig. 10.1 that you cannot view *comfortably* when reading this book from a distance of 60 cm will not be easily read when projected under the circumstances sketched above. When preparing slides on a computer with a screen width of 30 cm, you should assess the readability of the slides from a distance of 3 m, rather than from the distance of less than 1 m, used when working.

Again, slides or transparencies should support the speaker's story, never distract from it. Therefore, the material should be uncrowded, easy to read and understand, and pleasing to the eye. Crowding a slide with several figures, and text displayed in various font sizes and colors, serves only to challenge the listener. Don't force-feed the audience with an overflow of visual information. The top slide of Fig. 10.1 contains unnecessary detail. The mountains shown at the bottom serve no purpose other than to distract the audience. Likewise, the smoking chimpanzee, and the unreadable cartoon of cell-invasion by the influenza virus are equally irrelevant and distractive.[14] The clean, readable presentation of the numbers shown in the bottom slide of Fig. 10.1 is much more effective for conveying the message of this talk that the risks we run by daily activities, often taken voluntarily, are much larger than the risk we are exposed to from earthquakes.

Here is a clear *don't*. Do not use long sentences – and worse, paragraphs – on a slide. The audience that attempts to read them loses attention that should be devoted to you, the speaker, and your

---

[14] When showing these slides, we discovered to our surprise that not only children are amused by a smoking chimpanzee, so are graduate students and mature scientists. While humor can be useful at times for breaking the monotony of a presentation, unless close to the point of the message it can needlessly detract from the conveying of essential factual information.

message. The same holds for excessively long lists of bullet items.[15] Wordy slides force the audience to divide allegiance between reading the slides and listening to the speaker; again, slides should augment your message, not detract from it.

Color is helpful for making distinctions between different objects in a slide. Keep in mind, however, that multiple colors are distractive as well. We therefore recommend no gratuitous use of color; use no more colors than serve a functional purpose. Of importance, a significant number of men are red–green color blind. These men are unable to follow a speaker who states that "the difference between the green and the red curves indicates that . . ."[16]

LCD projection can strengthen a presentation and facilitate showing examples and illustrations in a convincing way. Remember, though, that the projection is no more than a tool for supporting a presentation; don't allow it to be a source of distraction from the message and you, the speaker. Your personal presence as speaker can be a valuable asset to delivering the message, while also making a personal impression on the audience. For this reason, we favor using LCD projection with moderation. For example, an effective way to start a presentation can be to start with a blank screen while speaking to the audience, before the lights are dimmed. Likewise, you can insert black slides in a presentation to draw attention away from the projection towards you while you are making a key point or transition, or while summarizing. (In PowerPoint, toggling the "b-key" makes the screen go black and back on again.) This convenient feature is especially useful during the question-and-answer part of a presentation.

More detailed advice on preparing slides can be found at numerous websites, such as

---

[15] Bullet slides, which should be included only for capturing key ideas, can benefit from deleting articles such as "the" and "a," as well as most verbs.

[16] Also, avoid use of dark colors for text and lines, such as red, against a dark background, such as black. While you might see no problem when you view your slide on your computer screen, the lines and text will disappear when projected on a conference-room screen. The same holds for poor choices such as light blue text on a dark blue background.

- http://www.cwp.mines.edu/~klarner/guidelines.html
- http://seg.org/publications/news/
  presentations01302008.shtml
- http://pages.cs.wisc.edu/~markhill/conference-talk.
  html

10.4.3 Delivering an oral presentation

Work on delivering your talk at a pace that is comfortable for both you and your audience. Nervousness and the concern for running out of time can make us speak too quickly, giving the audience less time to digest the material and causing listeners to tire quickly. Consciously setting a comfortable pace for yourself can reduce a feeling of nervousness, give you peace of mind, and aid in giving a presentation that is relaxed. Speakers are often unaware of speaking too rapidly. Again, the rehearsal with colleagues can give the helpful feedback for slowing down and timing your presentation.

Paradoxically, silences and pauses are an effective part of an oral presentation. A brief silence is the (un)spoken equivalent of the period sign at the end of a sentence. Many speakers omit these silences out of nervousness or the fear of running short of time. The result is that their presentation sounds like one long sentence. Silences play a larger role than simply that of separating sentences. When moving from one part of a presentation to the next, it is useful to insert a longer pause. This is the equivalent of the separation between chapters in a book.[17] Such a pause makes it clear to the audience that the presentation is moving into a new part, and also offers a moment for both you and your audience to relax. For some reason, some speakers often find it difficult to be silent when standing in front of an audience. A simple trick for inserting a pause is to count "one-two-three" (silently, of course). Being silent for an added second or two in front of an audience might appear like a century to the speaker, but for the audience it is

---

[17] Your talk is indeed a story, so it benefits from presentation that clearly conveys, through a slight additional pause, that a new chapter is about to begin.

a welcome pause in the flow of information. When asked a question there is no need to say "eeehh" or "uuhhh" when you need a moment of time to think. Simply be silent. Your audience will appreciate this and be less apt to interpret silly sounds, which serve no purpose, as a sign of incompetence.

Through eons of evolution, in common with that of other animals, humankind's impulse is to hide when confronted with a fearful situation. As a result, many speakers have a natural tendency to want to hide when confronted with an audience, surely conveying their nervousness.[18] In contrast, by being "out there," making full use of body language during our presentation, we convey confidence and a professional bearing.[19] *Don't underestimate how much and how effectively we can communicate through our body language.*

> *... no book can convey the special spirit and delicate peculiarities of its subject with that rapidity and certainty which attend on the sympathy of mind with mind, through the eyes, the look, the accent, and the manner in casual expressions thrown off at the moment, and the unstudied turns of familiar conversation,* in the 1851 essay "What is a university" by John Henry Newman (Cardinal Newman)
>
> *Huberman and Huberman, 1971.*

Body language includes our eyes and the eye contact that we make with the audience. A helpful practice during your presentation is to set your sight on and address your talk to a few selected individuals of the audience who appear particularly interested in hearing what you have to say.

---

[18] When overhead projectors were widely used, many speakers would hide behind the overhead projector. This illuminated their face from below, which gave an impression akin to that of children who, in the dark, shine their face from below with a flashlight. The resulting "nostril-illumination" conveys neither a pleasant nor confident impression.

[19] By having one of your rehearsals videotaped, you can develop your use of gestures (using hands) and movement (using legs). This practice of using gestures purposefully can also help you channel nervous energy that otherwise could work against you.

Speakers before large audiences have a large and growing array of technical audio/visual equipment at their command, LCD projector and computer, microphone, and laser pointer being the most frequently used to date. It is crucial to make sure that all the equipment works properly, and that you are familiar with its use. Many speakers think making sure that all the equipment is working properly is the responsibility of the organization that hosts the presentation. While, strictly speaking that might be so, when equipment fails, it reflects badly on only one person – the speaker. For this reason, it is much in your interest to be familiar with use of the equipment and to check well ahead of the presentation that everything works properly. Microphones, for example, might need to be activated with a switch; make sure you know where that switch is located. Also, knowing the location of the off/on button on a laser pointer avoids clumsy delays.

The laser pointer would seem to be a great aid for talks. When slides are well designed and are devoid of unnecessary clutter, one usually does not need a pointer at all. Our advice, therefore, is that if you must use a laser pointer at all, use it sparingly. Plan during rehearsals exactly when you will use the laser pointer and on which specific target on the slide you will direct the beam. When you are finished making your point, turn off the pointer. A laser beam butterflying across the slide or around the room is most distractive to an audience.[20] Also, be aware from which side the laser pointer shines its light; this avoids sending a laser beam into the eyes of the audience. It is amazing how many speakers complain that the laser pointer does not work while pointing the back-end of the laser pointer to the screen.

One further tip we offer pertains to the conclusion portion of your talk. Speakers, particularly those who are novices at giving presentations, have a tendency to wrap up their talks by extolling the virtues, but ignoring the disadvantages of, and what doesn't work

---

[20] Similarly, turn off the laser pointer while going over the content of word slides. Your audience is quite able to read the text, so they don't have to "follow the bouncing ball."

in, their new method. Since research developments often have some shortcomings, your standing as a scientist will be enhanced by having a reputation for sharing both the pros and cons in your findings.

For most conferences, time is allotted at the end of each presentation, unless the speaker runs over time, for members of the audience to ask questions. Questions establish a dialogue between you, the speaker, and the audience, offering the opportunity for clarification of points made in the talk. In addition, they give you feedback, letting you know how well those points have been understood. For these reasons, *welcome questions*. To help those in the audience who have not heard the question, we recommend that you repeat questions that have been asked. In addition, you might have misunderstood the question; repeating it can help avoid misunderstanding that leads you to give an answer that is unrelated to the question. You need not answer questions immediately. It is fine to be silent briefly while thinking about your answer. Some speakers are so keen to get their point across that they interrupt the person who asks the question. Aside from the chance that you might have missed the point of the question, an audience is likely to interpret this as a sign of rudeness or arrogance on your part.[21]

Having said this, sometimes the questioner can be aggressive or obnoxious. In that unfortunate situation, avoid becoming emotionally involved; pausing before answering gives time to think and can help you maintain a professional composure. Although it might not be easy when questions are asked aggressively, or when you are emotionally involved, always be respectful of questioners. A helpful idea is to communicate that you think the question is relevant: "this is a very interesting question" or "thank you for raising this point" or "you make an important point; please let me clarify." A polite tone for your answers will always be more favorably appreciated by the

---

[21] The website http://www.lbfdtraining.com/workshops/speakersbureau/resourcemanual3.html offers tips for new presenters who might find the question-and-answer session challenging, particularly when confronted with tough questions.

audience than one that comes across as arrogant, overly defensive, or sarcastic.[22]

10.5 THE SCIENTIFIC CONFERENCE

The scientific conference constitutes a key platform for exchanging the latest scientific results, often several months or even a year before they are published in the technical literature. In addition to being useful for keeping up with the work of others, the scientific meeting is a place for meeting with colleagues and having discussions about research. Epple (1997) gives useful information about organizing scientific meetings, including insight into how these meetings function.

Scientific conferences can be gigantic; it is not uncommon for more that 10,000 scientists to attend a meeting, and that the meeting offers more than 40 parallel sessions. With such a wealth of topics to choose from, it is generally difficult to choose which sessions to attend. Even with careful planning, it is impossible to attend all the presentations that appear to be interesting and relevant. Not surprisingly, many scientists, especially graduate students, feel lost at meetings of this size. It can be difficult to find persons you seek at large meetings, and the sheer size of these conferences creates a sense of anonymity that is not conducive for establishing contact with fellow scientists.

Despite all this, the contact you might establish with colleagues is probably the most useful aspect of a conference. This is the place to meet fellow scientists, to discuss your work, and to form personal ties with colleagues. How do you do this in a large mass of people? Sitting in the audience of sessions for the entire day is clearly not the best way

---

[22] One of our colleagues, a most polite individual who has an impaired hearing problem, was asked a question in an aggressive manner following her talk at a conference. She replied "Since I have a hearing problem, I cannot quite understand your question. Let me walk up to you, and perhaps you can repeat the question." The personal contact that resulted dissipated the aggression of the questioner, and a constructive public dialogue followed.

to meet others. A typically more fruitful use of your time is to spend some of that time outside the sessions. The hallways near the sessions of interest often are a convenient place for meeting others, including speakers, who are likely eager to strike up conversations about topics discussed in the sessions. Poster sessions are also good places to meet people because, there, scientists roam around from poster to poster seeking opportunities to talk with colleagues.

Scientific meetings often have social events that offer further opportunities to initiate contacts and exchange ideas with others. These include receptions sponsored by companies and larger universities. Also, special student receptions present useful occasions for graduate students to establish contact with students from other universities.[23] An open and outgoing attitude, with your antennae well extended so as to know when and where they take place, will enable you to take advantage of these social events.

The foregoing, however, does not mean that oral sessions are not useful. A session at a scientific conference can offer an excellent setting for becoming acquainted with a field of research. Many conferences offer sessions with the specific purpose of giving an overview of a certain field. The speakers in these sessions are usually invited, and the level of the presentations generally is high. These sessions offer a good opportunity to broaden your perspective on research in and beyond your field.

Apart from the exchange of scientific ideas, conferences are places to get to know colleagues. Often, when there is no organized evening program, the evening offers an opportunity to socialize and establish personal ties with colleagues. These ties help to create personal friendships within the scientific community. While this might be a valuable purpose in itself, these ties are also useful from

---

[23] As a graduate student, don't underestimate the importance of building good relations with other graduate students. They, rather than the scientists whom you might perceive as being *important*, are your colleagues of the future. More senior people, the authors of this book included, will have left the stage of research as your career reaches its peak.

a professional point of view, since many collaborations are driven by personal rapport between scientists. When going to a meeting with several people from your home institute, it might be tempting to hang out with these colleagues. A conference, however, is a place to meet scientists from *other* research groups; it can be helpful in the long run to exercise this opportunity. Conferences are also great places to meet potential employers in academia and industry. Employment opportunities are often driven by personal contact, so it can pay off to actively pursue personal contact with potential employers at a scientific meeting.

Perhaps paradoxically, the larger the meeting, the harder it is to meet and get to know people. For this reason, small meetings, such as workshops and summer schools, organized by many scientific organizations and with attendance less than about 150, can be much more useful for establishing contacts. For beginning graduate students, these smaller meetings are especially helpful for getting up to speed in a certain line of research and for establishing contact with both students and senior researchers from other institutions. Sometimes, these smaller meetings are held at remote locations that provide settings especially conducive for close discussion and interaction with other attendees, thus offering excellent opportunity for forming professional and personal ties with your peers.

During scientific conferences one usually meets many people, and numerous topics are discussed, many of which need follow-up action (e.g., sending a reprint or data, looking into a specific scientific question, or investigating a funding opportunity for joint research). The value of a conference can be greatly increased by indeed following up on such action items. Be aware that, unless you take notes, names and follow-up actions are easily forgotten after a couple of days, making it worthwhile to keep a small notebook or palm-size computer handy.

Don't be shy about establishing contact with fellow graduate students, more senior researchers, and potential employers. There is no harm, for example, in asking, after discussing science for a while, if a colleague you've newly met has no dinner plans and would like to go out for dinner together. Often, other people then tag along, creating an interesting mix of people who share a pleasant evening meal together.

Meeting people is one of the major spin-offs of scientific meetings, both large and small. To promote follow-up with those you meet, you will want to exchange contact information. For this reason, it is worthwhile to carry business cards, offer yours to others, and ask them for theirs. The cost of business cards is modest compared to the budget of most research groups, and their effectiveness in fostering and maintaining contact usually makes this a worthwhile financial investment. These cards serve not only for exchanging contact information; they can also be used as handy reminders, for example, by writing a note such as "I look forward to receiving your reprint" on the back of the card. Be aware that, in some cultures, a business card is customarily treated as a treasured gift. Putting a business card in your back pocket without looking at it can be taken as an insult. Instead, accept it with two hands, look at it carefully, and treat it as a treasure. (And, indeed, the future contact might turn out to be a treasure.)

10.6 CONCLUDING NOTE: YOU, THE AUDIENCE

Above, we have emphasized that effective communication requires that the sender, i.e., the speaker or writer, be attuned to the needs of the receiver. As a member of the audience, it is equally important that you make a concentrated effort to understand the message of the speaker or writer. Again, most people absorb only a fraction of the words that are spoken, especially in a time-limited oral presentation. Minimizing distractions while listening is the least you can do to maximize your understanding of the speaker's message. Of equal value is that you ask questions of the speaker whenever the arguments are not clear. Besides helping you fix ideas you've heard, this gives the speaker the feedback needed to pitch her story somewhat differently.

Likewise, when reading a book or article, it takes the dedication of the reader to gain thorough understanding of the content. Although skimming the pages diagonally is a valuable way to decide on whether or not a given paper is worth your delving into at greater depth, it does not do justice to the effort of the writer to convey a message. It takes two to communicate.

11 Publishing a paper

Publications are the primary means for distributing, establishing, and archiving scientific results. The decision to hire or promote somebody is often based to a large extent on the number and quality of publications that the individual has written. Because, typically, the number of copies of a journal article that are printed is orders of magnitude larger than the number of thesis copies made, papers in technical journals are of much larger value to the scientific community than are theses. Because most graduate students must prepare and defend a thesis, the best of both worlds for them exists in graduate programs that both encourage students to publish their research work during the course of their studies and allow them to use their published or submitted papers, perhaps in an adapted form, as chapters in their thesis. For all the above reasons, scientific publications are of great importance. In this chapter we suggest questions to contemplate prior to writing a paper, steps to take during the submission and review process, and actions to consider while the paper is in press and afterward.

11.1 BEFORE YOU START WRITING

Before writing a manuscript you need to decide in which journal you intend to publish your work. The choice of journal for publication can be of crucial importance. Among other reasons, this decision can influence the tone and content of the paper, its length, and the format you use.[1] Several factors enter into this decision. Many journals, for instance, have specific restrictions on the maximum length of papers,

[1] Different journals have readership with highly differing interests and perspective. As discussed in the previous chapter, of prime importance in communicating ideas is that you know and understand your audience. Therefore, tone and content of a paper should be tailored to your understanding of the readership expected for the journal under consideration.

the number of references cited, and the scope of the work that they consider. The choice of a European journal versus an American journal dictates whether you write your article in American English or in the Queen's English. The following questions can be relevant when choosing a journal in which to publish.

- *Do you want to publish in the journal that has the largest readership or one that has the strongest reputation in general, or in your field?* Some journals have much larger influence than do others because they have a larger distribution, because, for whatever reason, their reputation is particularly strong within a particular scientific community, or because they are highly selective about the papers that they choose to publish. The *impact factor* of a journal is a quantitative measure of the influence of the journal in the scientific community. This factor, computed by the organization Thomson Scientific, is the average number of citations to papers in a given journal within two years of publication. Thomson Scientific's list of impact factors, which can be viewed online, offers the scientific community a clear sense of which journals are most influential. The general scientific journals *Science* and *Nature* have by far the largest readership in the broad scientific community and the largest impact factor.[2] A low impact factor for a journal, however, does not necessarily mean that the journal is not influential. Some journals, for example, mostly serve an industrial research community. Since researchers in industry are likely to publish less often than are their peers in academia, this can lead to a low impact factor, although the work published in that journal could well be much read and highly influential. In practice, it is more difficult to get a paper published in the top journals than in journals with lesser reputation in their field. It takes experience to decide on the most appropriate journal to which to submit your work for publication.

---

[2] Impact factors of many journals are surprisingly low. In many research fields, the top journals have an impact factor of around 5. Numerous journals have an impact factor less than 1, meaning that, *on average*, papers in those journals are cited less than once in the two years following publication! *Science* and *Nature* have impact factors in the range 20–25.

- *Who do you want to reach?* It could be that your work is of special interest to researchers in a specific part of the scientific community. Not only might you have a better chance of getting your paper published in a journal that is highly respected in a specific field rather than in the broad area of science, you could well prefer to catch the eyes and ears especially of researchers in that particular community. So, do you want to reach your fellow specialists, a broad range of scientists, or the general public? You might also want to consider the geographic distribution of the journals in which you intend to publish. Scientists in the USA sometimes tend to ignore journals published in other countries, even when these journals are of the highest quality. Seek the journal that best can reach your target group.

- *Do you want to have your results publish rapidly?* When engaged in highly competitive research, it can be of crucial importance to publish results quickly before a competing group possibly claims similar results. Some journals specialize in rapid dissemination of results. Apart from *Science* and *Nature*, numerous specialized journals aim at rapid publication. The names of these journals usually contains the word "Letters" as in *Physics Review Letters* or *Biomedical Letters*. Likewise, some journals with conventional turnaround times to publication have a special category of paper for quick-turnaround treatment, usually also called *letters*. Whether in letters-journals or in journals that have the special category, letters, papers published as letters usually have restrictions on their length.

- *How well does the journal handle the review process?* As we discuss in Section 11.2, papers submitted to scientific journals are reviewed by experts in the field to ensure the quality, clarity, and relevance of the published work. The speed of the review process, quality of the reviews, and professional care and timeliness of the editorial board that handles submitted papers vary among scientific journals. Choosing a journal with a reputation for an efficient, professional, and fair review process can help avoid delays and aggravation.

- *What is the cost of publication?* Perhaps surprisingly, many journals charge authors for publishing papers. The so-called *page charges* can be considerable and do vary widely among journals. For some journals, page charges are mandatory, while for others they might be voluntary, and some have no page charges at all. Color figures in a publication usually cost considerably more than do black-and-white ones; when planning to use color figures, it can be advantageous to choose a journal that charges modest costs for these figures. Therefore, know the page charges and other costs of publishing in a journal before deciding to submit a paper there so you can decide whether the page charges, if any, are an expense worth bearing.

- *Do you want to support for-profit journals and publishers?* Scientific papers are written by scientists, and editorial boards of journals are staffed by scientists, mostly on a voluntary basis. The scientists often pay page charges, and scientific libraries purchase journals, often for a hefty price. Some journals are published by for-profit publishers whose aim is to make a profit; one might question if this agrees with the spirit of the work done on a volunteer basis by editors, reviewers, and authors. Many journals, though, are published without the aim to make a profit. Those journals are usually published by professional organizations or by non-profit publishers, such as Cambridge University Press. The Institute of Physics (IOP) contributes to education and development by making its publications available in electronic form and free of charge to many developing and low-income countries. All things being equal, you might prefer to send your work to journals that are issued by not-for-profit organizations. This helps control costs of scientific journals and promotes dissemination of research results to underprivileged groups.

- *Do you want to use open-access or open-source publications?* Modern information technology makes new types of publications possible. *Open-access* publications are distributed in electronic form without the involvement of a publisher. One of the earlier databases of such publications is arXiv.org[3] with so-called *eprints* in physics, mathematics, computer science, quantitative biology, and statistics.

---

[3] http://arxiv.org.

Open-access publication is rapidly evolving, and a Google-search using "eprint" gives pointers to a large number of initiatives. *Open-source* publications go even further by treating the publication as an evolving electronic document. A popular example of an open-source document is Wikipedia[4], the online encyclopedia to which anybody can contribute. In open-source scientific documents, other researchers usually have access to the data and software used for the analysis and can contribute their findings and interpretation to the electronic document (Murray-Rust, 2008). Open-source publishing increases sharing and promotes discussion and dialogue among researchers. The Public Knowledge Project[5] is one of the numerous organizations devoted to open-source publishing. Both open-access and open-source publications are publicly available, and thus contribute to the free dissemination of scientific results. This is an example of how information technology serves in the education and democratization of the poor in the world, as described by Friedman (2007). Consider using these alternative ways of publishing your work when the opportunity exists in your field.

Some of these considerations can be conflicting. It takes experience to make the best decision on where to submit your work. Colleagues can be helpful as sounding boards regarding the choice of journal. In any case, consider the choice of journal carefully before submitting (even before writing) a paper.

Once you have decided to which journal you will submit your work, you likely will have to tailor some details of writing style to the requirements of the journal. Most journals have specific rules that manuscripts must satisfy upon submission, rules that range from critical issues such as the length and scope of manuscipts to more trivial conventions such as the mandatory use of SI-units. These rules are described in the *Instructions for Authors* that usually can be found in the January issue of the journal or on the website of the

[4] http://wikipedia.org.
[5] http://pkp.sfu.ca.

journal. Because the publication of a paper can be delayed unnecessarily if you ignore a journal's requirements, be sure to comply with its instructions.

11.2 SUBMISSION AND REVIEW

So let's assume you've followed the guidelines of the intended journal. Manuscripts are usually submitted electronically, although some journals still require paper copies. Many allow you to submit your manuscript to your choice of one of their editors. In that situation, it is worthwhile to look over the list of editors and make a careful choice as to whom you send your manuscript. When possible, send your manuscript to the editor who works in a field that is close to yours so that your work will be well understood and be treated most professionally. Do you know any of these editors personally?[6] If so, are they fair-minded and objective? Do they have enough backbone to overrule the comments of a reviewer when that reviewer is wrong? Do they have a reputation for handling papers without undue time delay? Choosing the editor carefully can help avoid complications and delays later in the publication process. Many journals invite authors to suggest names of reviewers. Suggesting colleagues who are knowledgeable, fair, prompt, and not driven by territorial instincts can help avoid major problems and delays. In short, take advantage of windows of opportunity offered by the journal that can speed up and otherwise aid the review process.

Once the editor has received your manuscript, you should expect confirmation of receipt by the journal and usually a submission number for future correspondence about the paper as well. Most journals now use electronic submission and almost immediately send the confirmation notice by email. For the rare journal that requires a paper submission these days, notification of receipt should

---

[6] While you cannot be expected to know the editors early in your career, you might find this situation changed later on. Also, ask your adviser or other colleague if they can recommend one of the editors for that journal.

arrive within about two weeks. If you have not received confirmation in a reasonable amount of time, check with the journal because manuscripts are sometimes lost or mislaid.

The editor sends your manuscript to several reviewers (usually three or four). These individuals, chosen for their knowledge in the field, read your manuscript and advise the editor on the scientific quality and originality of your work, the clarity of presentation, and the suitability of the submitted work for publication in the journal. Usually the editor follows the advice of these reviewers. A good editor, however, will make a clear decision when the reviews are at odds, possibly because one is blatantly wrong. Also, it unfortunately can happen – infrequently, however – that a review is biased because the reviewer has an interest in blocking your work from publication.

The review stage can take anywhere from a few weeks to half a year or more. You should receive the reviews in a reasonable amount of time; for most journals this is about three to four months. If you have not received the reviews by that time, don't hesitate to contact the editor and inquire about the status of the review process. Too often, reviewers who have agreed to review a paper are simply too busy to be timely in reading the manuscript and writing the review. They then need to be prompted by the editor to complete the review. Although it is the rare exception, a reviewer might stall intentionally in order to delay the publication of a paper. Usually, however, it is safe to assume that a tardy reviewer is simply overwhelmed by her work. After a few months, be persistent in asking the editor at monthly intervals for the reviews of your manuscript. As elsewhere in life, it is common courtesy to remain polite in doing so; polite inquiries are much more effective than are irritated reproaches. Moreover, the editor is the last person you want to irritate, even when you are angry or upset.

Once the reviews arrive, you will usually be asked to make modifications to your paper. Sometimes, the reviewers have not understood the message correctly. In that event, you should not blame the reviewers. They are the experts in fields for which the article was originally intended. If these experts did not understand your paper, likely you have not phrased your words carefully enough, and a thoughtful rewording can turn around their understanding. As author, you are the individual responsible for altering the text to clarify points that reviewers misunderstood.

This, however, does not mean that you have to honor all the comments and suggestions of the reviewers and editor. They could simply be wrong, and you disagree with them for valid reasons. Then, do not hesitate to follow your own course; you need not incorporate all requested changes in the revised manuscript. When submitting a revised version to the editor, it is customary to write a letter in which you state which changes you have made in the revised version of the manuscript. If you have not incorporated some of the items suggested or required by the editor or reviewers, then carefully explain in your letter why you have taken these decisions. Don't play a hide-and-seek game of ignoring suggestions with which you disagree. Most editors are reasonable individuals who will respect your decision if your reasoning is sound, and if it is presented in a manner they can follow.

There is another good reason for carefully explaining to the editor which changes you have made and which not in the revision process. Editors are scientists who typically serve as volunteers on an editorial board in addition to their regular job; these individuals are busy. If it takes excessively long for an editor to understand how an author has handled the comments of reviewers, then that editor is likely to send the revised manuscript once again to the reviewers. This causes delays, and introduces the risk that these reviewers raise new issues that were not part of their original review. These are complications that could have been avoided if the author had provided the editor with a clear description of the revisions made in response to the comments of reviewers. You help the editor; the editor helps you.

It sometimes happens that a manuscript is rejected or receives harsh criticism for reasons that seem unfair. In such situations, you might judge that starting a discussion with the editor will not be helpful or worthwhile. The best strategy in that circumstance can be to withdraw the paper formally from that journal and submit it else-where. It has happened to us that a paper was in the review process for more than a year, after which the reviews were highly negative, and (in our subjective opinion) strongly biased and unfair. This prompted the editor to require a major revision that would have entailed rewriting most of the paper. After withdrawing the manuscript, we submitted it to a competing journal that accepted it on the basis of the highly positive reviews it received. Fighting a battle that is already lost has no value. Instead, seek an alternative route to publication.

The peer-review process has been highly valuable to the advancement and integrity of the sciences. It should be considered of great value to the individual author as well. Having successfully passed the review process, a paper has been individually judged to be of merit to the relevant scientific community. Moreover, peer review offers the particular value that the paper is likely improved by the process, made more clear not only to the selected reviewers, but, as important, for readers of the paper once it has been published. The journal peer-review process is an extension of a practice that we rec-ommend to all authors – in-house review of their paper by colleagues prior to submittal for publication. You should regard all reviewers of your papers – your colleagues as well as those participating in the journal peer-review process – as friends whose constructive criticism benefits both you and your paper.

11.3 AFTER PUBLICATION

In the absence of complications, a paper usually is published within about 1 to 1½ years after the date of submission of the original manuscript to the journal. Some journals specialize in rapid publica-tion and reduce this time down to several months. In either case, you shouldn't expect that, once your paper has been published, everybody who could be interested in your work will read the paper and cite it in their work. As discussed in Section 9.2, keeping up with the literature is difficult for anyone because of the large amount of papers published. For this reason, your peers might not be aware of your paper once it is in press. If you want them to read it, you probably will have to place your paper under their noses. In the past century this would have been done by sending reprints to colleagues. Now you can also send your work around by email, preferably in pdf-format. (The suffix *pdf* stands for "portable data format," a format that can be read on a wide variety of computer systems with the free program *Acrobat Reader*.) Don't ignore the part you can play in the dissemination process by distributing your paper to the group of readers for whom it is most relevant. Be aware, though, that most journals and publishers require the author to sign a transfer-of-copyright statement. The copyright restrictions to which the author has agreed could limit the distribution of copies of your work on paper or in electronic form, so it is prudent to check with the journal to learn what is allowed. Most often, the restrictions will not apply to papers you provide to colleagues.

12 Time management

Time is all you have. And you may find one day that you have less than you think.

Pausch, 2008*

"I just don't have enough time to do all that I need to do." This complaint has become almost the mantra of life in modern society. The fretting expressed here invariably leads to a state of physical and emotional stress that is often detrimental to the well-being of both our professional and personal lives. Key words in the mantra are *[not] enough time*, *need*, and *just*. The problem is not that there is not enough time – the time allotted to any individual in life is fixed (but unknown in advance). Therefore, most truly, we each have just the right amount of time for whatever it is that we choose to do in our lives. Rather than there not being enough time for all that we *need* to do, the problem is that we *want* to do too many things in a given amount of time. It's a matter of *choice*. The word *just* in the mantra suggests that we have no choice in the matter. The key message in this chapter, however, is that we *do* have that choice.

12.1 SETTING PRIORITIES

This brings us to the central point of time management. Given the finite amount of time available, we have essentially two ways to reduce the feeling of having insufficient time – by choosing our activities carefully and by working efficiently. Starting with the first one, in general, we have many activities from which to choose. During your student career, you have courses to take, qualifying and comprehensive exams to prepare for, thesis research to pursue, literature searches to extend the breadth and depth of your scientific capabilities, social and cultural life to pursue, and on and on. The tugs in these many important directions do not lessen with time, nor does the number of directions decrease as you move through your career, whether it be in industry or academia. In the academic environment, one usually has one or more research projects on which to work, papers to write, scientific literature to follow, classes to teach, students who need supervision, faculty meetings to attend, equipment and computers to purchase and maintain, and research proposals to write, not to mention numerous little chores that command attention.

No wonder many of us feel under time pressure. In addition to these tasks, many scientists voluntarily take on additional ones such as serving on editorial boards or advisory panels, visiting schools for outreach activities, and reviewing research proposals or manuscripts. This is more than any one individual can realistically do, even in a workweek that is excessively long. By saying "yes" to every activity in which we could possibly be engaged, we can drive ourselves crazy in a very short time. The key is to make choices that are based on thoughtfully arranged priority for those activities. We often find it difficult to say "no" to a request because we are eager to please, because we underestimate the time involved in the new task, because of guilt or fear,[1] or simply because we are interested in taking on that new task.

Here is a different perspective on the hard choices that need to be made while setting priorities. Suppose that you are extremely busy, and somebody asks you to take on yet another task; do you say "yes" or "no?" A key point to realize is that when we are overloaded and say "yes" to yet some other task or project, we *are* saying "no" to something else. That something else can be another project, it can be our health or peace of mind, it can be the amount of sleep we get, it could be time spent with friends or family, or it might be a general reduction in the care and quality of the work carried out. It can be illuminating to analyze what alternative choices you are saying "no" to when you say "yes" to a new task or project. A clear understanding of how a

---

[1] For some people, guilt or fear are unfortunately all-too-important factors in decision-making. These emotions are negative ones, and the choices resulting from them often lead to comparably negative results.

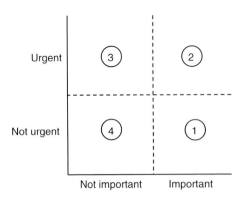

FIG. 12.1. *A division of activities.*

new task will influence other aspects of your work and personal life might make it easier to actively choose and say "yes" to the activities that indeed deserve the highest priority. Looking at the choice to be made in this way makes it easier to deal with the feelings of guilt or inadequacy that we might otherwise experience by saying "no."

Stephen Covey (1990) provides valuable insight on time management and priority-setting in his wonderful book *The Seven Habits of Highly Effective People.* Figure 12.1 is adapted from his book. In general, activities can be either important or not. Importance is, of course, a property on a sliding scale, but for the sake of argument we simply use the labels "important" and "not important" along the horizontal axis of the figure. Activities can also be divided roughly into those that are urgent, and those that are not. Urgency, also simplified from its true sliding scale, is depicted along the vertical axis of Fig. 12.1. According to this figure, our activities fall into one of four categories: those that are important and not urgent (quadrant 1), those that are both important and urgent (quadrant 2), those that are urgent and not important (quadrant 3), and those that are neither urgent nor important (quadrant 4).

In general, many of us tend to give priority to activities that are urgent over those that are not. The urgency of something tends to prompt us to action, and it is easy to confuse urgency with importance. The statement that "a memo should be submitted before the end of the afternoon" is easily confused with the statement that "this memo is important," because the matter discussed in the memo might be of little relevance. It is useful to make a thoughtful distinction between urgency and importance. One might wonder why we would spend *any* time on matters that are not important other than because we confuse urgency with importance. We do this for a number of reasons: we might happen to enjoy doing a particular activity that is not important; we might procrastinate in working on issues of higher priority because we don't know where to start; we might be unaware that an activity is indeed not important; or we might be driven by guilt or by the pressure of our superiors to do things that we do not consider important. Analyze the reasons for spending time on non-important issues, and then make a decision as to whether or not to continue giving those issues your attention.

Many of us tend to respond to the urgent matters. The result is that activities that are important, but not urgent (quadrant 1), don't get the attention they deserve. In a research environment, examples of activities in this category include maintenance of research infrastructure, acquiring new skills, submitting long-term research proposals, building good personal relations with colleagues, and aspects of our personal well-being that include health, family, and friendships. It takes careful prioritizing and planning to ensure that sufficient attention is given to the activities of quadrant 1.

Even with thoughtful attention given to the distinction between the urgent and the important, the choices that we make can lead to a schedule that is overfull with activities. The struggle to balance professional life and personal life is particularly difficult. Being a successful researcher requires a significant investment of time. Meanwhile, we have a personal life to live that could, for example, involve making and maintaining friendships, satisfying cultural and intellectual interests, playing sports, engaging in spiritual or religious activities, doing volunteer work, raising a family, and aiding with the concerns and health of friends and relatives. Choosing professional activities that lead to a workweek of 70 hours or more is bound to lead to conflict between our professional life and our personal life. This means that our choice of activities must involve not only the management of our time at work, but also a conscious choice for a balance between personal and professional activities. This is often the hardest choice to make because it involves balancing activities that are part of such disparate aspects of our life.

It can be enlightening to look at the way you actually do spend your time, and who decides which activities you engage in. Do you set your priorities, or do others make decisions about how you spend your time? Who is in the driver's seat for putting together your agenda? Ideally, we make the choices ourselves and use our time in conformity with the priorities that we attach to the various activities. In practice, this is not so simple as it sounds, since we work together with others who also have ideas about how we should spend our time. All this talk of making decisions and setting priorities can seem of limited value in a hierarchical situation where we are simply told to do something. Such a situation is more likely to arise in a business environment than in research, but it definitely does happen that researchers are simply told to carry out certain tasks, perhaps more commonly so in industrial research groups than in academia.

In the hierarchical relationship between graduate student and adviser, the student might feel that he has little choice but to say "yes."[2] Even then, the student usually has *some* room for influencing the course of action and various ways of exerting such influence other than by saying "no." For example, it may be possible to offer alternatives such as "can we push back the date of completion to this task?" or "what is the risk of taking this alternative approach?" Such alternatives often have the best chance of being accepted when the person who gives the directions feels that those are her ideas rather than yours. Let others invent your solution! When confronted with yet another task in a situation of overload, it can be useful to ask your

---

[2] The healthiest of relationships between adviser and student are far from being demandingly hierarchical.

superior how you can make room for that new task. ("In order to spend sufficient time on this new task, which activity should I push down on the priority list?") Such decisions are most effective when they are jointly taken.

The way in which we spend our time cannot and should not be driven solely by our own priorities and wishes. The goal of prioritizing is not to be independent of others. Working or interacting with others makes the setting of priorities and the planning of our activities a matter of give-and-take. For this reason, Covey (1990) speaks about a state of *interdependence* rather than one of *independence*.

12.2 USING TIME EFFECTIVELY

Scheduling one's time is an important aspect of time management. This is especially so for researchers because original and creative research is possible only when one is fully involved in research. It is the rare individual who can do effective research in half-hour segments. In such a short time span, one can do a routine task such as entering data points or making a figure, but it is virtually impossible to be truly creative unless one is completely absorbed in the research. Therefore, it is essential to plan the workweek in such a way that it contains blocks of time that can be fully devoted to research. Time management thus involves choosing not only *which* activities we want to work on, but also *when* we want to work on them.

The electronic calendar is a helpful tool for planning time effectively. A particular advantage of the electronic calendar is that it can be made readable for others, thereby allowing communication of our schedule to others. An electronic calendar, for example, allows a secretary to set up appointments, while being informed at all times about the current schedule. Keep in mind, though, that no amount of software or clever gadgets can solve the problem of time management when not combined with a conscious choice as to where to put priorities.

The main steps in good time management are thus setting priorities of our activities and planning the moment for these activities with thought and care. These two steps remove much of the feeling of not having enough time. A number of others steps, however, can also assist in managing time effectively.

- *Make a list of things to do.* The stress that we experience over the feeling of having too much to do is caused in part by an apprehension over forgetting some tasks of importance. Making a "to-do" list takes away this concern and can also help in setting priorities, which provides the most solid basis for ordering tasks. The to-do list can simply be written on a piece of paper. Alternatively, index cards and an electronic task planner offer tools for ease in shuffling activities according to their priority. A to-do list also provides the opportunity for the feeling of accomplishment that comes when crossing a completed activity off the list.
- *Aim to arrange your schedule in such a way that you do activities at a time of the day for which you are at your best for working on them.* Some people feel energized and creative in the morning, but tired in the afternoon. For them, it is a good idea to do research in the morning and to carry out routine activities later in the day. Analyze the time of the day at which you are at your best for a given activity, and plan your time accordingly.
- *Analyze how you spend your time.* Despite all efforts at careful time planning, we often spend our time differently from ways that we had planned. Often, we are not even aware of the extent to which we get sidetracked during a day. A simple way of monitoring this is to occasionally[3] write down all activities carried out in a day and compare this list with the intended schedule. The discrepancies can be revealing. Being aware of them makes it possible to plan time more realistically. It can also be illuminating to use a diagram such as Fig. 12.1 and place all activities that you carried out during a day or week in the quadrants where they fit best. This gives clarity on whether you have focused on the important activities rather than the

---

[3] Only occasionally. You don't want this monitoring to become a major activity.

urgent ones. Also pay attention to activities that you did not carry out, but that do deserve attention. You could add them to the diagram in a contrasting color to indicate that you currently are not giving attention to these important activities. Perhaps this might prompt you to spend your time differently.

- *When you are in the position to do so, more likely later in your career than while you're a student, ask a secretary to handle your agenda.* Planning a meeting with several people, all of whom have busy schedules, is a time-consuming and often frustrating task. It is normal in a professional environment that the secretarial staff take care of this activity. Delegating the time scheduling can be a huge time-saver, especially when an electronic calendar system is shared in common for establishing and communicating the schedule. Note that this delegating does not necessarily imply a loss in control of your schedule. For example, by blocking out certain parts of the week for research, writing, or other activities that require much concentration, one ensures that those parts of the week will not be invaded by other plans.

- *Delegate tasks whenever possible.* Again, this option will become more available once you have completed your student career. Delegating work relieves the workload and makes it easier to focus on the more important tasks. Delegating work expresses confidence in the person asked to carry out the task; this expression of confidence is often appreciated despite the work that comes with it. Most researchers in academia find it difficult to delegate, perhaps because delegation involves a certain loss of control. When others are given work that you could have done yourself, it is unavoidable that sometimes mistakes are made by those others. Accepting that such mistakes will happen should be understood as a fact of life for delegated tasks as long as the mistakes don't occur too often. The cost of mistakes that will be made occasionally in a delegated task should be balanced against the alternative of not delegating at all, with the result of having to deal with a workload that is unacceptably high, and thereby making mistakes yourself!

- *Avoid unnecessary distractions.* Electronic means of communication, such as email, cell-phone, and, to a lesser extent, the normal telephone, offer a rapid and often highly convenient means of communication. The downside of these tools is that they can be the source of a continuous stream of distraction.[4] It is, unfortunately, too common nowadays that participants in meetings are checking cell-phones or palm-top computers for messages; no wonder many of us feel hurried and unfocused! Not used with care, email and other such means of communication can divert attention to the extent that moments of concentration disappear from the normal workday. Song *et al.* (2007) provide useful advice on the efficient use of email. When starting the day with reading of email, the temptation is large to let the activities of that day be driven by the incoming email. A problem with email is that, when used improperly, it tends to spur a continuous flow of activities that are perceived as urgent, i.e., those that fit in quadrants 2 and 3 of Fig. 12.1. Some of those are important, but many are not, and it is easy to spend much of the workday on activities that sidetrack us from those that we consider most important. A good way to reduce the distraction of email is to read it only at the end of the day. In that way it is easier to make conscious choices about the email to which to react. Similarly, unless they are truly needed, cell-phones can be switched off or (better yet) smashed.
- *Take good care of yourself.* People generally work most effectively when they are healthy and well rested. Occasionally, one hears stories of people who can do with only five hours of sleep per night. Both of us have tried reducing the amount of time sleeping; these attempts have typically ended in ill-temper at best and errors in judgement at worst. Our bodies have certain needs, among which are enough sleep and sufficient physical exercise. Taking care of your physical well-being gives your mind a healthy vehicle in which to operate. The mind also has its spiritual needs, whether recognized or not. Hard as it may be to

---

[4] The phrase *internet snacking* was recently coined to indicate that users of the internet use this medium in an increasingly fleeting and distracted way.

give shape to those needs, it pays off to give attention to this aspect of our well-being as well. To put the pressure to produce in perspective, we offer advice that we received from our friend and colleague, Rodney Calvert, to whom we dedicated this book.

Take care, and do not work too much – it dulls the man.

He was Chief Scientist of one of the major oil companies and one of the most productive and centered researchers in our field.

Note to conclude this chapter. Much of what we've written in this section is an example of "do as I say, not as I do." We speak from experience in saying that effective time management is not easily accomplished; lapses can arise with regularity. The largest sources of such lapses arise from failure to recognize explicitly that the choices are *ours to make consciously.* The key steps listed above, consciously followed, are of immense help in staying on track. Of course, it can happen that you are one of those people who wish to do it all – all the tasks deemed necessary to succeed in research and scholarly pursuit, being a helpful friend to students and colleagues in need, model parent, voracious reader on all subjects, amateur astronomer, pursuer of all that the outdoors has to offer, exercise and sports enthusiast, community activist, avocational musician, volunteer firefighter, furniture maker, ... If you feel you must try to do everything, good luck in getting all of your priorities neatly arranged and pursued in rational order. Our primary recommendation in that situation is that you be aware that these are choices that you've made and that you use this awareness to minimize your inevitable sense of frustration. With your long list of things you feel you *must* do, you don't need to add fretting and frustration to the list.

13 Writing proposals

We are all agreed that your theory is crazy. The question that divides us is whether it is crazy enough to have a chance of being correct.

Niels Bohr

Writing research proposals is an integral, although perhaps the least favorite, part of doing science. In an academic environment, the financial resources for carrying out the research come, to a large extent, from external sources. Similarly, in an industrial environment, research proposals are often the means by which a researcher must convey to management the value of supporting a project. Proposals thus are essential for securing research funding, but they are important for other reasons as well. Having research proposals funded shows that your research stands up to peer review, that it has the qualities judged worthy of receiving financial support, and that you are able to communicate research plans in a compelling way. It thus constitutes a sign of professional expertise in job applications and often is considered of critical importance in decisions regarding tenure and promotion. Research grants can sometimes be transferred along with the individual when switching to a new university. Bringing in already-existing research funds can positively influence the decision to make a job offer to the holder of those grants. Research grants thus not only provide the financial means to carry out research, they increase the market value of scientists applying for a job or seeking tenure or promotion. Much has been written on the preparation and submission of proposals. We offer suggestions for further reading on this topic in Appendix A.

13.1 WHO FUNDS RESEARCH?

Proposals can be submitted to any of a number of different types of sponsoring organization, including the following.

- *Government agencies.* Most countries have organizations that allocate federal or national funds to research groups. In the USA, the National Science Foundation (NSF) is a general sponsor of research; it has a counterpart in most countries. In addition to the general science foundation, most governments have other funding agencies that specialize in support of particular types of research. Examples in the USA include the National Institutes of Health (NIH), a major sponsor of biomedical research, and the Department of Energy (DOE), which sponsors research related to energy and environmental issues.

- *Industry.* Because of the potential spin-off from academic research, industry also sponsors research at universities. The research sponsored by industry often is focused on applied work with specific deliverables, but industry often sponsors more risk-prone "game-changer" academically oriented research as well.

- *Private foundations* also provide funding for research and education. In the USA, for example, the W.M. Keck Foundation supports students and faculty, and finances research infrastructure such as the Keck telescope in Hawaii. These foundations vary greatly in size, the larger ones having grant programs that invite submission of proposals.

- *Academic institution grants.* Universities often provide financial support as a tool to steer new directions in research or education, sometimes using seed money for new initiatives and to attract larger grants. Often, one can apply for this support by submitting a proposal to an internal grant program of the institution.

- *Private donors.* Research sometimes is funded by support from private donors. Usually this support is provided to universities rather than individual research groups, and it is most often used for highly visible purposes such as new buildings or the foundation of new research or teaching facilities. The size of donations can be considerable. For example, in 2005 John A. and Katherine G. Jackson donated more than $300 million to the University of Texas at Austin to found the Jackson School of Geosciences.

The mechanisms by which these organizations distribute resources differ substantially. Government agencies, academic institutions, and, to a lesser extent, private foundations decide how to allocate resources based on their evaluation of proposals submitted to them. Often, the proposals are submitted to programs focused on either a specific discipline or a particular research opportunity, sometimes spurred by political developments. The North Atlantic Treaty Organization (NATO), for example, for many years sponsored joint research between NATO countries and those of the former Soviet Union and its allies. The research programs to which proposals can be submitted are usually announced with a Request For Proposals (RFP),[1] which provides a description of the type of proposals that can be submitted and guidelines that must be followed when submitting a proposal. These requests for proposals usually are posted on the websites of funding agencies. With many funding agencies, one can sign up for email alerts for RFPs in chosen areas of research.

Industry-sponsored research and private donations are often driven by personal contacts rather than by requests for proposals. Fellow researchers from universities and industry might focus on common research questions that lead to industry-sponsored academic research. Industry support for academic research can arise for yet other reasons, for example, in the hope of recruiting graduate students to become future employees. In that case the education of young professionals, rather than the research itself, is the prime reason for providing financial support. Private sponsors often make donations based on a personal connection with the university or other institution to which the donation is made, such as when the donor is an alumnus. In such situations, a proposal might still need to be submitted, but it often serves more as a tool to specify and document the plans that are made rather than as an entry into an open competition for research funds. Research proposals can thus play any of several different roles depending on the type of organization to which they are submitted. They always nevertheless contain an account of

---

[1] For some reason, funding agencies love abbreviations. An abbreviation such as RFP has by now received the status of a new noun in research land.

the proposed research, the reasons why the research is important, expected results, deliverables, and a timetable.

Writing proposals is a time-consuming and energy-consuming activity that is especially frustrating when the proposed research does not get funded[2] or when the size of the grant is judged barely worth the effort of having gone to the trouble of writing the proposal. Writing proposals nevertheless can be an activity of direct value to the quality of the research itself. It forces us to articulate the current state of research, define innovative research questions that push the field forward, develop a strategy to address these questions, and understand what resources would be required. Writing a proposal thus serves a role in planning research, even when the proposal doesn't subsequently receive funding.

13.2 THE CORE CONTENT OF A PROPOSAL

As a proposer it is critically important that you consider what you would want to have explained before you gave someone else $100,000. I am regularly astonished about investigators/reviewers who trash other proposals for being sloppy in their wording or for having poorly defined objectives, who then basically say "Give me money; I am great" in their own proposals. Don't think that program directors don't notice it.

Nick Woodward, US Department of Energy, 2007, personal communication

In this section we focus on the core part of the research proposal – the planned research. Other parts of a proposal, including the summary, budget, reference list, and biographic sketches of the investigators, are discussed in Section 13.3. Minimally, five ingredients are necessary for a proposal to be successful:

- a high quality of the submitted proposal,
- a research area in need of progress, and reasons why the proposed work needs to be addressed now,

---

[2] Depending on the research area, funding organization, political and economic climate, and other factors, the competition for winning research grants might be such that only a small percentage of proposals submitted to a given funding agency (as low as 10%) are awarded grants.

- an interesting research question/objective in that area,
- a research program at a funding agency into which the proposal fits well, and
- clearly defined deliverables.

What does it mean that a research proposal be of a high quality? The proposal must be well written, with a research plan that is sound and developed in sufficient detail to be convincing. Although it might appear to be an extraneous cosmetic detail, it is also important that the proposal "looks good." Enhancing the readability of the proposal, a factor that can help shorten the reading time for the reviewer, will make a favorable impression on reviewers and on members of selection committees. Often, an upper limit is imposed on the number of pages allowed in a proposal. Don't attempt to squeeze too much into a proposal by reducing the font size that you use. This makes the proposal harder to read and usually creates an unfavorable impression on the readers of your proposal.

In order to appreciate this last point, it is useful to know how most proposals are evaluated. Most funding agencies send proposals to independent reviewers; usually these are peers in the scientific community. A panel of experts reads the proposals and reviews, and advises the funding agency on the ranking of the proposals. Program managers of the funding agency then make a final ranking based on this advice. The members of the panel receive all the proposals and all reviews. Often, this is more than 1000 pages of written material. The panel members are supposed to read all this material, but in practice they simply don't have the time to do so. Moreover, the program managers of the funding agency typically are each responsible for several programs. This makes it even harder for them to actually read proposals. In practice, many of the proposals are read "diagonally," or at most only the summary and brief other parts are read. It is therefore important that the proposal be a pleasure to read and that the relevant information be easily extracted by the reviewers and panel members. For these reasons a research proposal must be no longer or more verbose than necessary; a good proposal is written in an incisive and readable style.

The great paradox is that most scientists who submit proposals aim for completeness spelled out in excruciating detail, whereas program managers and panel members simply look to find the basic information in the proposal. They look for answers to the following questions: What are the objectives of the project? What is the true innovation, the novel aspects of the proposed work that will help overcome current limitations in our knowledge? What is the research methodology? Does this research group have a reasonable chance of success in carrying out the proposed work? Why is this the right moment to fund this work? What specifically is being requested? What is the track record of the investigators? What benefits does the proposed research offer to the science or technology, to the research community, or to society at large? Aiming to include fine detail can actually make it more difficult for the panel members to extract the essential information within their limited time available. Writing the proposal with this idea in mind can help in making the proposal more attractive and professional.

Clearly, without a good research question and a promise of innovation that pushes the field forward, a proposal will have a poor chance of receiving funding. The basic scientific question that underlies the proposed research forms the backbone of a proposal. Formulating an interesting research question, however, is not enough. Most funding agencies will support only those proposals that address topics that are aligned with the goals of certain well-defined programs. These programs usually focus on a particular discipline (such as "Stem Cell Research"), but they can also focus on specific themes of societal relevance, driven perhaps by a political goal (such as "Joint Research Between the USA and Japan on Earthquake Disaster Mitigation"). For a research proposal to have a good chance at being successful, an appropriate match must exist between the research question and the goals of the research program formulated in the RFP to which the proposal is responding. By investigating with care the opportunities offered by funding agencies, you can aim for an optimal match between your research ambitions and the current interests of the funding agencies. The internet is the most efficient tool, and the preferred way, for obtaining information on programs that are inviting proposals.

A proposal is all the more compelling when the deliverables of the project are explicit and clearly stated. Besides making the proposed research tangible, this aids in making a professional impression on reviewers, panel members, and program managers. Nick Woodward, program manager at the US Department of Energy, suggests the following content for the core of the "perfect proposal."[3] It would have a length of 15 pages along the following lines.

- *Page 1.* Executive summary (not an abstract) that immediately allows the program director to identify (1) the objective, (2) the people, (3) the techniques to be used, and (4) the type of reviewer expertise needed.
- *Pages 2–5.* Concise description of the "state of the art" in the subject area of the proposal.
- *Pages 6–8.* Concise and polite description of what is lacking in or wrong with the current "state of the art." Identify a vision of the future for the research area and what are the major gaps, even if they will not all be addressed in this proposal.
- *Pages 9-13.* Concise descriptions of (1) the highest-priority objectives, and (2) the innovative approaches the investigator will take to overcome the limitations in the current state of the art. Justify why not all gaps will be addressed immediately.
- *Page 14.* Identify the cast of characters and the unique skills that each brings to the project.
- *Page 15.* Discuss the potential scientific ramifications of the research. This always involves understanding the products of the research well enough to be able to predict how it will work (1) in the future or (2) in various possible spheres of application.

Proposals to an industrial sponsor often are much shorter and less detailed than those submitted to, for example, a national funding

---

[3] He notes that he has never seen the perfect proposal.

agency. The reason for this difference is that proposals to industrial sponsors are usually driven by personal contacts. Before actually writing the proposal, much of the groundwork for the proposal already has been done through discussions between colleagues in industry and academia. Furthermore, an industrial proposal is usually not submitted as an entry into a competition based on a request for proposals. Sometimes, proposals to industrial sponsors have, as their primary purpose, the need to convince management of those organizations that it is worthwhile to support the research. Given the limited time that managers devote to such issues, these proposals usually are short. (Managers are busy and are more interested in relevance and deliverables than in scientific detail.) We have been asked to submit to industrial sponsors proposals that do not exceed two pages. Proposals for industrial sponsors usually have a much stronger emphasis on deliverables and timelines than do proposals of a more academic character.

13.3 THE OTHER PARTS OF A PROPOSAL

In addition to the project description, proposals are typically requested to contain an abstract, budget, bio-sketches of the principal investigators, and a list of references. As mentioned in the previous section, panel members and program managers are faced with the daunting task of reading thousands of pages when handling a single round of proposals. It is unrealistic to expect that they will read and digest such a vast amount of material. It usually is a requirement that an abstract or summary be included as part of the proposal, and, for many reviewers, this might be the only part of the proposal that they read. Therefore, although the temptation is large to consider the abstract as an afterthought after the main part of the proposal has been written, this approach does not do justice to the importance of the abstract. The abstract must capture succinctly the essence of the proposed work by describing the main research question, the research approach, the innovative aspects of the proposed research, and the impact of that work, and it must be written in a way that is engaging and sparks the interest of the reader. All this must be crammed into a limited amount of text. For these reasons, writing a good abstract is difficult and requires much thought and good writing skill.

In the budget, one specifies how the funds requested in the proposal will be used. The budget typically consists of personnel costs (such as tuition and stipend for graduate students and summer salary for faculty[4]), the purchase of equipment and materials, travel expenses, publication costs, other administrative expenses, and overhead. Overhead is a fee that most universities charge to cover the expenses of running the physical and administrative infrastructure of the university. The overhead charge can be considerable; an overhead rate as much as 50% is common. For such an overhead rate, a proposal with $200 000 direct costs actually costs the funding agency $300 0000.[5] Research proposals offer an opportunity to cover expenses that could be difficult to cover otherwise. Examples of such expenses include the cost of travel, page charges (see Section 11.1), and administrative costs (e.g., copying, telephone). Don't miss the opportunity to cover those expenses. Also, because the budget must meet the administrative criteria and policies of the host institution, it is advisable – even required – to use the help offered by the office of research management (or whatever fancy name might be used) of your institution for putting the budget together.

The background and reputation of the investigators is an important consideration for funding agencies when evaluating proposals. The RFP therefore includes a request for a brief description of the background of the investigators. This so-called *bio-sketch* is an abbreviated version of a curriculum vitae that summarizes the education and employment history of the investigator, as well as highlights of

---

[4] At most American universities, faculty have appointments for nine months per year. Research grants can be a source of income for the remaining months.

[5] Because of the income generated by overhead charges, universities value the successful generation of research funds. This is one of the reasons why the ability to write successful proposals (and, of course, to conduct quality research) can be an important part of promotion and tenure decisions.

her publication list. In Section 15.2.2, we offer general advice on how to write a curriculum vitae. In practice, the funding agency gives fairly strict guidelines for the content and length of the required bio-sketch. While staying within these guidelines, don't be overly modest. This biographical information is part of a proposal that will be judged on a competitive basis, so don't hesitate to toot your horn.

As with any scientific manuscript, a proposal should contain an appropriate bibliographical reference list. This list provides background material for reviewers, panel members, and program managers that, one hopes, shows the investigator's awareness of the current state of research. The reference list offers an opportunity to display your experience in the field of the proposed research by citing your own work, but the list should not be overly biased toward your own work. Remember that the proposal will be reviewed by a number of knowledgeable colleagues. These reviewers usually work in the same field, and ignoring their contribution in the reference list could lead to less-favorable recommendations to the funding agency. As with every part of the proposal, keep in mind those who will be reviewing it.

13.4 WRITING AND SUBMITTING THE PROPOSAL

Before writing the proposal, it is often helpful to obtain some background information about the RFP or possible future RFPs from within the funding agency. The most direct and best way to get critical information and insight about funding opportunities is to talk with people at the funding agency. Find out who is the program manager or grant officer of the program to which you are considering submitting a proposal, and give that person a phone call. Even though these people are often busy, they are usually helpful in providing information. In fact, one aspect of their job is to communicate effectively with the scientific community and assist in the targeted submission of relevant proposals of high quality. Since personal contacts are beneficial, you can even consider visiting a funding agency in order to get to know the people involved on a personal basis. (And, in the process, they of course also get to know you.) You might wonder why it is advantageous to talk with the people at the funding agencies when so much written information is available. Much important information, however, is not included in the request for proposals. Consider the following questions you might pose in a meeting.

- What fraction of submitted proposals can be expected to be funded? This question is of importance to the decision about whether or not you want to invest the time and energy in writing proposals directed to a given program. Some RFPs don't request a full proposal, but require a brief pre-proposal instead. Based on the pre-proposals, selected investigators are invited to submit a full proposal. This approach has the advantage that one spends time in writing a full proposal only when there is a reasonable chance, say 50%, that the proposal will be funded.
- In practice, what is the average or typical amount of funding given to each proposal? The request for proposals often gives only the total amount of funding available for a program and the maximum amount of funding that can be requested, but programs often have a rule of thumb concerning the amount of funding that has a good chance to be approved for a proposed research activity. You should ensure that the amount requested is considered realistic by the funding agency.
- Apart from the official evaluation criteria, are there other considerations or informal criteria that will be taken into account in the decisions of the funding agency and its advisors? Such considerations can be specific (e.g., they have reasons to prefer research area X over research area Y), but they can also be highly vague (the panel "likes" a particular type of research).
- What does the program manager think of the question underlying your proposal? Many program managers are willing to look at an outline of a proposal before you actually write the full proposal. Their comments often are extremely useful.

Therefore, don't hesitate to make personal contact with a key person – possibly the program manager – in the funding agency. This could require some perseverance. Once you finally talk with the right person, however, you will probably discover a willingness to help you where possible and within reasonable bounds. By doing so you also might have established a useful contact for future opportunities. Be aware that program managers need to be careful not to engage in a conflict of interest. They might not be able to provide every piece of information that they have or that you request. Also, for good reasons, they are unable and unwilling to accept any gifts or favors. Buying a program manager a drink at a scientific meeting is already considered to be off-limits. Do not tell the program director what he or she should have their agency do; this is not in their power, and hearing such statements is annoying to them.

Once you think you have found a program that fits your proposed research well, pause before starting to write the proposal. Each program of any given funding agency will have specific criteria that proposals must meet. You will want to be aware of these criteria and be sure your proposal is written in such a way that it matches them as closely as possible. Some of the criteria are objective and require only trivial effort for you to satisfy them. When, for example, it is stated that the project description can be at most 10 pages, you know that it cannot be 11 pages; don't exceed the page limit. Other criteria, however, can be more vague. As an example, the requirement that the proposed research must have "Social Relevance" cannot be translated into simple rules to follow. Because your proposal should optimally match the evaluation criteria of the funding agency, it is clearly advantageous to have a full understanding of all of them. The best way to achieve this is to make a list of all the criteria that your proposal should meet. You can then use this list as a guideline while designing and writing the proposal; moreover, you can also use it as a checklist once you have written a draft of the proposal.

Once you have decided on a research question and a program to which you will submit your proposal, and have made the list of evaluation criteria, next make an outline for the proposal. In Section 10.3 we give suggestions aimed at helping with the writing process. For now, we reiterate the important point that you start with a rough outline and then make the outline progressively more detailed until you arrive at the point where you know well what you want to say. Only then are you in the position to start writing. Most important: once you have written the proposal, give it to a few colleagues and ask them for their feedback. Their comments are crucial because it's human nature that, when we write a proposal or article, we tend to be myopic in sighting our own errors and shortcomings. This is even more so when we are under time-pressure and cannot afford the luxury of putting the work away for a while, to return to it later with a fresh mind.

Many funding agencies and journals invite suggestions of names for reviewers of your proposal. This provides a valuable opportunity. The objective is not that you channel your work into the hands of a "buddy" who will help you in an unfair way; in the long run (and even in the short run) this is a poor strategy. You do, however, want your work to be reviewed by those who you know to be knowledgeable and fair-minded. When your proposal or article is judged by a reviewer who is not knowledgeable about the area of your proposed research, the review could well turn out unfavorable, primarily because the reviewer did not understand the material well enough to appreciate it. Also, unfortunately, not everybody has an open mind and a constructive attitude. This can be aggravated by territorial instincts when your work competes with the work of the reviewer. Suggest knowledgeable and fair-minded reviewers whenever possible.

We have argued that personal advice by program managers of funding agencies can be crucial for the success of a research proposal. The personal relationship with the research sponsor is of even more value when requesting support from industrial sponsors. These sponsors usually don't issue requests for proposals; proposals submitted to industry typically are the result of personal contact between industrial and academic researchers. For a proposal to be successful, the contact person in industry must fulfill two criteria. First, that person must be knowledgeable in the research and must have a passion for the work that matches that of her academic colleague. Without such expertise, common ground, and passion, there is little chance that the industrial colleague will choose to push for sponsored research. But this is not enough; not every researcher in industry has sufficient clout to convince management to allocate resources to the proposed research. To convince management, it takes professional expertise, the ability to build a strong case, a fighting spirit, and sometimes seniority. It is therefore not sufficient just to have an enthusiastic and capable colleague in industry with whom to collaborate. When that person lacks the required personal skills or status within the organization, it is unlikely that enthusiastic plans for research support will actually materialize.

14 The scientific career

The career in science for which graduate school prepares you could take many different possible shapes. Depending on your field, opportunities exist for stimulating and rewarding work in either industry or academia. In this chapter we outline different types of scientific careers, the sorts of choices to be made, and considerations worth keeping in mind when making these choices. Be mindful, however, that none of the choices carves the direction of your career in stone; that direction can and quite likely will change over the course of time. Even though changes are possible later in your career, greater effort is often required to make the transition later rather than earlier on. Planning ahead can help create favorable conditions, but don't expect that you can plan for every eventuality. Expect the unexpected in life if for no other reason than that you could well discover that your outlook, ambitions, and circumstances change with time.

Because the range of career possibilities available is so wide, varied, and fundamentally personal, the general topic of career choice is complex. Moreover, types of career choices change with time as society changes. Recent and up-to-date information on the scientific career can be found in the journal *Science Careers*, issued by the American Association for the Advancement of Science (AAAS). The American Association of University Professors (AAUP) has a website[1] that covers important career issues in higher education, and Appendix A lists further reading on the topic of the scientific career.

14.1 THE ACADEMIC CAREER

Upon completion of graduate studies, the first-order decision to make is whether to continue a career in academia or to work in industry.

---

[1] http://www.aaup.org/AAUP/issues/.

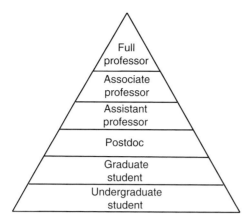

FIG. 14.1. Hierarchical structure typical of the academic community in the USA. Elsewhere, such as in Europe and Japan, the definitions of academic rank differ, but the structure largely has its counterpart stratification and pyramidal shape.

Not every field of science offers that choice, but in practice young scientists with a Ph.D. or M.Sc. degree generally do have the choice. In this and the following section we characterize research careers in academia and industry.

For the academic career, it is useful to understand the structure of the academic community as sketched in Fig. 14.1. After completing an undergraduate degree and graduate education, the most likely entry into an academic career is as a postdoctoral fellow, usually abbreviated as *postdoc*. At most universities in the USA, professor appointments are made at three levels in increasing hierarchical status: assistant professor, associate professor, and full professor. This difference in rank is usually reflected in salary level. The various academic ranks and their meaning differ from one country to another. In many European countries, for example, the academic community has a particularly strong sense of hierarchy, with professors in the lower ranks typically reporting to those with a higher rank. In the United States, the different ranks seldom reflect a chain of command, but serve more to indicate and reward different levels of experience. The pyramidal shape of the figure shows that the number of scientists decreases with increasing rank (although the width is not scaled in any meaningful way).[2]

Since the reader is likely to be either a student or at a stage beyond, we discuss only positions of postdoc and higher. The appointment of postdocs is almost always on a temporary basis, while appointment at the professor level most often aims to be permanent. Postdoc positions are temporary (*non-tenured*) appointments, whereas the permanent positions usually are referred to as *tenured* ones. Some professors are partially, or completely, funded by research grants; positions funded by grants are called *soft-money* positions. For a full-time soft-money position, the university provides neither salary nor the job security of tenure, but offers workspace and the use of its research infrastructure, usually in exchange for overhead charged on the research grants that fund the position. Although such positions offer no assured long-term job security, because the funding comes from external sources they entail no obligations such as teaching and university-service responsibilities other than those imposed by the funding agency. This gives researchers on soft money considerable freedom from university responsibilities, again at the expense of little or no job security.

Conventionally, the path toward a tenure-track position typically starts with postdoc positions, temporary research positions usually funded through external grants. The first postdoc position often is exciting. You move to another university, often in another country, perhaps another continent. You meet new people and are exposed to new developments. This also is a time in a research career with a minimum of constraints imposed by others; in graduate school there is the adviser to whom to report, while in later stages of the academic career pressures imposed by the employer and other job responsibilities mount. In practice, young scientists often have one or more postdoc positions before entering a position that leads to

---

[2] We do not intend to suggest that the pyramidal shape reflects a *food chain*, wherein those at higher ranks devour those at lower, nor does it imply that rank equates in any way with personal superiority.

tenure at a university. Postdoc positions become less attractive with time; the novelty wears off, and moving could become increasingly difficult, particularly for the family that the postdoc might have by that time. Moreover, a drawback of postdoc positions over the long term is instability of the soft-money support; soft-money funding for salary is available only as long as research grants are in place. This adds up to being in the *postdoc trap*; it is highly advisable to move beyond a postdoc situation as soon as is realistic to take up a more stable position to your liking. Because of the ever-increasing number of postdoctoral fellows and the larger number of publications required for a faculty appointment than in the past, many researchers spend an increasing amount of time working as a postdoc in order to build a sufficiently strong portfolio of publications. As described bitterly by Griffin (2002), postdoc positions offer a number of significant disadvantages relative to faculty positions – the absence of career structure, limited salary level, lack of possibilities for promotion, and unsatisfactory retirement benefits. To this, add the instability in one's personal life associated with frequent job changes.

Despite the drawbacks of a temporary position in science, the postdoc position usually plays an important role early in the career development of scientists. A postdoc position helps one acquire the experience and visibility needed to eventually obtain a permanent position in academia, and it offers the freedom to set up one's own line of research. It is through publication of quality research that an individual develops the reputation needed for a faculty position.

Most positions in academia involve both research and teaching. Postdoc positions, however, often are limited to research only, with the drawback that the experience obtained as a postdoc is lopsided towards research. It is valuable to acquire some teaching experience during your postdoc period so that you have a more balanced background when applying for a more long-term academic position. Since your postdoc position probably is funded from external grants, you might need to be insistent about your desire to do some teaching because your employer will expect you to work only on the research for which the grant was awarded. Likewise, to help you prepare for an academic career, it is much to your advantage to be an outgoing participant in various activities within your department that go beyond just the research. Be a resource for graduate (and undergraduate) students, and, to the extent available in a reasonable workweek, offer to teach portions of courses and to present seminars.

In many countries, independent research fellowships are available that support young scientists. These fellowships are typically awarded on a competitive basis and often are highly prestigious. To receive such a fellowship often is a huge career boost, not only because it creates the opportunity to set up an independent line of research, but also because these fellowships usually are viewed as a sign of exceptional research talent. The science foundations of many countries offer prestigious awards to promising young scientists that come with the financial support for starting a research group.

Almost no universities will offer a permanent position to someone immediately after completing a postdoc position. (You, in fact, might well choose to avoid those that will offer you a permanent position at that stage because this could be an indication that the university has a problem attracting high-quality scientists.) Instead, universities offer *tenure-track* positions. At some European universities this often is a two-year trial period. At most American universities this is a period of, at most, about six years, within which you have to prove your capabilities for producing high-quality research and teaching, and, often particularly important, for attracting research funding. If you pass these tests in accordance with the standards of the university, your temporary position will probably be converted into a permanent one, a tenured appointment. Those for whom tenure is denied generally lose their tenure-track appointment and must move to a position elsewhere.[3]

---

[3] Under certain circumstances, e.g., when a candidate had applied for tenure at a date prior to the final year by which she must apply, she might have the opportunity to strengthen her case for tenure and re-apply after a year or so.

Before accepting a tenure-track position, it is good to investigate the criteria that the university uses in making decisions on tenure. Be aware that the official criteria could well differ from those used in practice. A good way to gain that awareness is to talk with colleagues who presently have tenure-track positions, or with those who have recently gone through the tenure-track stage at the university where you apply – both those who obtained tenure and those who did not. Useful advice on tenure evaluation is given in the report "Good practice in tenure evaluation," which resulted from a *Joint Project of The American Council on Education, The American Assocation of University Professors, and United Educators Insurance Risk Retention Group* (2000).

In the United States, the tenure-track process is an essential step towards a tenured academic position. Other countries have other mechanisms to evaluate the academic stature of an individual before offering a permanent position, or making a promotion to the highest academic ranks. For example, the *Habilitation*, as used in Germany, involves writing a second thesis beyond the Ph.D. thesis, giving several seminars that are judged from a scientific and educational point of view, and showing academic leadership as evidenced by publications, obtained research grants, and classes taught.

While the process for seeking and attaining tenure in the USA[4] can vary greatly among universities, the period prior to securing tenure can be demanding and intensive, especially in so-called *research universities*.[5] The process for seeking tenure in any given

[4] Although the meaning of tenure can differ greatly from one country to another, at some stage in a career anywhere in academia some formal evaluation process for promotion into or through the system will be required.

[5] Universities in the USA can broadly be divided into those classified as primarily teaching universities and those that are known as, or consider themselves to be, research universities. The spectrum in-between is continuous, with plenty of research done in teaching universities, and vice versa. Many outstanding universities have students at only the undergraduate level, but even those universities can have faculty who do highly credible research. Universities that are consistently listed among the top, e.g., Ivy League universities, CalTech, MIT, Stanford, and most state universities, among others, are considered research universities.

university is usually spelled out in the *Faculty Handbook* for that university. The approval process proceeds in stages that include evaluation of the application for tenure by tenure committees in the home department and in the university, as well as by the department chairman and someone higher in the university administration.

The candidate must create a dossier that shows evidence of excellence in research and teaching. A dossier could typically contain (1) the CV, including the full record of publications, both peer reviewed and not, presentations at professional society meetings, workshops, and symposia, (2) all courses taught, and student evaluations of those courses, (3) committee and other service to the home department, parent school, and university, (4) professional activities, such as service on organizing committees for symposia, and workshops, and on journal editorial boards and other committees of professional societies, (5) honors and awards, (6) invited presentations, and visits invited by other institutions, (7) number of successful grants and amount of acquired research funding, and (8) letters of recommendation by noted scientists in their field.

Tenure committees in various institutions struggle with the decision on the relative importance to give to a candidate's strengths in teaching and research. Before choosing to apply to a given university, it is good for you to gain an understanding of the relative emphasis that the institution places on these two qualities.

In highly supportive departments, a worthy candidate can generally count on the help of champions among the department's tenured faculty in putting together the case for tenure. In the most supportive of situations, that help starts virtually the day that the tenure-track individual arrives on campus.

As sketched in Fig. 14.1, the number of people engaged at a given level in academic careers decreases significantly with increasing level of seniority. Some leave the scientific career because of lack of adequate career options, but there might well be positive reasons for leaving an academic career path. Some researchers decide to use their expertise to carry out research in industry or government laboratories because of the type of research carried out in those laboratories, job security, higher financial compensation, or societal relevance. Others choose to pursue a career other than research, for example, in business. Preston (2006) and Robbins-Roth (2006) give advice to those switching from an academic career to another career path.

14.2 COMPARISON WITH THE INDUSTRIAL CAREER

Much lore exists about large distinctions between academic and industrial careers. We have heard of "academic freedom" that may be found in universities, but which is absent in industry.[6] Likewise, we often hear of the distinction between so-called *pure* research pursued in academia, as opposed to applied research in industry. Such distinctions are blurred at best and, more often, non-existent. Historically, depending on the company and discipline, considerable freedom to pursue research often has existed in industry, without the pressure to publish or perish. Likewise, some companies in various industries have offered and continue to offer great freedom to pursue wide-ranging research that provides relatively little short-term reward for the company. Moreover, funding agencies increasingly lean toward support of university research that is of a practical, relatively low-risk nature. At the same time, market-driven pressures do push companies in industry toward more-applied research and development, on the basis that they will produce product within an ever-shorter time period.

The point here is that, while you may have a leaning toward one or the other – industry or academia – on the basis of some preconceptions, it could be well worth your while to look closely and openly into the real opportunities that best suit your interests and inclinations. Not all companies and all industries are alike, and university research and teaching environments can vary considerably. Small "high-tech" companies offer an excitement that is often not found in large corporations, while large companies with outstanding

---

[6] Conversely, the pressures of "publish or perish" in academia led one industry colleague to make reference to "academic tyranny."

218 THE SCIENTIFIC CAREER

research staffs offer opportunities for interaction with a wonderful team of strong scientists. Likewise, in a small "teaching" university one can find the low-pressure environment that suits a temperament for doing well-paced, careful research, whereas in a large university one has the opportunity to be surrounded by other high-energy, high-productivity researchers. Moreover, the large university has the advantage of having associates nearby with whom to collaborate in efforts to obtain funding. No single prescription for a career fits all.

A career within industry is harder to describe than one in academia because the variety of industrial careers is so much larger. Some companies offer an inspiring, almost academic, research environment (with regular seminars and invited speakers, but without faculty meetings) where people stay in research for extended periods. Other companies expect their scientific professionals to leave research after a certain period of time and take a position in management or operations, or perhaps they expect them to go into a non-research (but still highly technical) position at the outset. Some companies transfer their employees regularly to different locations, while others offer more stability in work location. Because of the personal ramifications of moving, it is important to know the policy of a potential employer. Learn what are the typical career paths for scientists in companies to which you apply. Again, discrepancies can exist between the official story told by the company and what actually takes place in practice. Talk with some of the old hands in the company as well as with people who have left the organization. The likelihood of a lifelong career with the same company or even in the same scientific discipline has been diminishing in recent decades, and the trend likely will continue. A plan based on a long career with a single industrial employer, although not out of the question, therefore could be unrealistic.

While individuals with the same general sorts of personality traits choose to work in either academia or industry, we might mention one distinction that can be highlighted regarding preference for one work environment versus the other. There is a tendency for those who choose to work in academia to function more as individuals, whereas, in industry, teamwork is especially valued. As mentioned above, interdisciplinary research in academia has great merit for both the individual and the research. The desired collaborations, however, are not enforced, and the degree of cooperation and interaction is as variable as are individual personalities. In contrast, research in industry is generally purposefully founded on teamwork and generally populated by individuals who especially value working together closely. As is true in all endeavors in life, however, the spectra in both environments are continuous and fully populated. Toward which end of the spectrum do you lean?

Most universities care far less about the age of members of its faculty than about their faculties' research and teaching skills. In industry, where most companies have more or less well-defined career paths for their employees, career steps can be closely linked to age. As one grows older, a scientist often becomes less valued as a researcher by industry, unless she brings specific needed expertise to the company. This suggests that it might be easier at a later age to move from industry to academia than the other way around. An industrial background can, however, make one attractive for a university because of the expertise and contacts acquired over a career, whereas an academic background is of relevance for industry only when it comes with specific scientific expertise that industry needs. A move in either direction is quite possible for younger individuals, but, in either case, only once they have established a strong reputation in their field. Perhaps an overriding factor is the economic climate of the moment. If an industry is expanding rapidly, the move from academia to industry can be as easy as slipping into a vacuum. In contrast, it is difficult to imagine large numbers of job openings in academia in the foreseeable future.[7]

---

[7] Researchers in industry tend not to publish so often as those in academia. Therefore, opportunities to move from industry to academia will be available mostly to those who have seen and acted on the benefit of publishing even when this was not valued by their employer.

Again, making a career change is possible only when your capabilities are attractive for a new employer. For this reason, it is important to build your market value as a scientist. Two worthwhile steps are to continue learning and to continue to expand your expertise over time, and a critical factor is the publication record. In judging job applications, most universities place large weight on your publication record, even though many search committees will deny that this is one of their primary selection criteria. Therefore, regardless of whether you work in industry or academia, in order to maintain your marketability for any change in employment, it is essential that you publish your work in refereed journals. The phrase *publish or perish* may be an exaggeration, but your market value as a scientist in academia is measured to a large extent by the number and quality of your (more recent) publications in relation to your age. Most companies in industry, however, do not place particularly high value on publication within the company (although publications count for much when a company is evaluating an applicant who has been with another employer). Many companies, unfortunately, discourage publication by their professionals, either through out-and-out denial of requests to publish or through a policy that puts up so many hurdles that the net result amounts to prior restraint. Therefore, for the sake of your future career (which could include a change of companies or a change of careers to academia), it is in your interest when considering a job with a new employer to learn about the company's publication policy. This information is likely best obtained from potential peer employees rather than from management.

The above comparison between the career in industry and that in academia has ignored that in government laboratories. The research done in government laboratories has much in common with that in academia, and the job application processes are similar as well. Likewise, both share the tradition of having less emphasis on the applied, although grant pressure has trended toward research that is more applied. Largely, research in national laboratories is funded by a grant application process similar to that used to support much academic research. Perhaps the largest difference is that teaching is not a requirement in the national laboratories.

14.3 SWITCHING FIELDS: BENEFITS AND PITFALLS

Almost always the men who achieve fundamental inventions of a new paradigm have been either very young or very new to the field whose paradigm they change. And perhaps that point need not have been made explicit, for obviously these are the men who, being little committed by prior practice or to the traditional rules of normal science, are particularly likely to see that those rules no longer define a playable game and to conceive another set that can replace them.

Kuhn, 1962[*]

A career in science can take on different forms depending on the breadth of your interests. Some scientists are super-specialists who are leading experts in a limited area of science, whereas others have much broader, but less deep, expertise.[8] Whether you want to be a specialist or a generalist depends to a large extent on your nature and inclinations. All things being equal, however, making changes in career, field, or focus allows you to build up a broad expertise, which has distinct advantages.[9]

Each field in science has its standard practices and beliefs, some of which can go to the extreme of amounting to dogma. Upon entering a new field, you have not yet been indoctrinated with the notions that pervade that field. This makes it easier to have a fresh mind that is open to new ideas and approaches. This mindset provides you with the opportunity for creativity that might have become diminished for more experienced researchers in that field. As expressed by our colleague Rutt Bridges:

In the beginner's mind there are many possibilities, but in the expert's there are few.

---

[*] Reprinted with permission of the University of Chicago Press.
[8] Breadth and depth are, however, not necessarily mutually exclusive.
[9] A tongue-in-cheek description of the scientific expert is given by Oxman *et al.* (2004).

Of course, you don't bring the experience, expertise, and reputation of old hands in that field; here, your openness to learning from colleagues who are willing to share their expertise and experience will greatly benefit your research.

Switching fields offers fresh opportunities for creative ideas at the outset in the new field. At the same time, as a newcomer, your ideas might well be the most naïve. There is nothing wrong with coming up with outrageous ideas that don't work (naïvity can be good), as long as you can identify them early as likely being unproductive. (Recall Section 7.4; make your mistakes quickly, but also recognize them quickly.) Besides the possibility of enhancing your creativity, switching fields can be productive for another reason. Often, ideas or techniques can be transplanted from one field to another area of research. Switching fields therefore provides you with a more varied toolbox for doing research, and can bring to the new field valuable cross-fertilization. In the process, however, avoid the temptation to project that, with your outside background, you are superman bringing the ready expertise to solve their long-standing problems. It could well be that the researchers in your new organization have considerable experience with intricacies of their problems of which you are unaware.

There is yet another reason why we often do our most original work when we start working on a new problem. Our mental view is unobstructed because it is not yet cluttered by the details of the problem. Those who have been working on the problem extensively might have lost that breadth of outlook and become bogged down in detail, making it more difficult to develop an efficient research strategy.

In many areas of research, some of the most exciting developments take place at the interface of different fields. Examples include quantum computing, the role of biology in the geosciences, and computer science in handling the massive data sets needed for unraveling the human genome. Work at the interface of different fields can be effective when researchers from a given field have the expertise and willingness to communicate with colleagues from other fields.

Switching fields takes a certain boldness but helps build up the expertise and mentality to work effectively with colleagues from other fields of research (Snieder, 2000).

This section has presented a view biased toward the development of scientists with breadth of expertise. Such an open and wide-ranging career does not fit everybody. Some prefer to continually refine and elevate their expertise in a specialized discipline. Such an approach works best when you actively choose to rejuvenate your skills and outlook over time so as to avoid becoming stagnant in research. This can be done equally well within the framework of a career as either a super-specialist or a generalist.

We have argued here that switching fields can be important. Interdisciplinary research currently is popular, and indeed many new fields of research have opened up because of the collaboration of researchers from different fields. There is a danger, however, that *interdisciplinary* research degrades into *non-disciplinary* research. The key in effective interdisciplinary research is that two or more researchers, who have mastered their own disciplines, collaborate to create a new line of research in which the whole is more than the sum of its parts. This works, however, only when the researchers involved indeed bring strong expertise in their field of research to the party. Interdisciplinary research can be ineffective when researchers who are broad, but not specialized in any field, collaborate; they lack the new expertise to bring to the problem.

14.4 AS YOU TAKE ON MORE RESPONSIBILITIES

A career in industry or academia that starts out in science often takes on a different flavor over the years as one accumulates more responsibilities. In practice, many scientists acquire management tasks over the years. In an industrial environment these tasks are clearly labeled as such, whereas in an academic environment they are usually hidden under important-sounding names such as "chairman of the such-and-so committee" or "head of the blah-blah department." Despite trendy notions about multi-tasking, we can devote our time productively and effectively to only one activity at a time. The added responsibility usually comes at the expense of our own research or teaching. Depending on your taste, skills, and ego, such a change in responsibilities can be a welcome shift in activities or an annoying distraction from research or teaching.

This new phase in your career has both upsides and downsides. On the upside, your capacity to do research generally increases as you build up a research group. In practice, many scientists lose interest over time in the nitty-gritty details of research and develop more interest in dealing with the broad picture of research. Typically, in such a situation, your productivity, as measured in number of co-authored publications, increases as your research group expands, but your skills for doing the detailed aspects of research may suffer. Expect the likelihood of this trend, and decide early on whether or not this is something you desire.

Taking on progressively more management tasks over time also has its downsides. Most scientists enter a career in science because of their passion for science. Over the years, however, they might discover that they have so many new responsibilities that they can no longer find time to satisfy that passion; moreover, they lack the time to practice and dig into scientific topics in depth, and lose the ability to follow the types of pursuit that prompted them to enter a scientific career. This tends to push these individuals away from their scientific work to other tasks for which, unfortunately, they lack the necessary skills. Making matters worse, many scientists succumb to *Peter's principle*, which states that *employees within an organization will advance to their highest level of competence and then be promoted to and remain at a level at which they are incompetent,*[10] an unfortunate outcome for both the individual and the organization.

Without proper care, organizational work has a tendency to grow. This is an example of Parkinson's law, which states that *work expands so as to fill the time available for its completion* (Parkinson,

---

[10] *American Heritage Dictionary of the English Language,* Fourth Edition, 2000.

1958). Without putting on the brakes, committee work and other management activities can easily expand to a degree where research is marginalized. The tragedy is that being a good scientist is no guarantee for having good managerial skills. Most scientists with administrative duties at universities were selected for their research skills rather than for their competence as managers or administrators. Without the proper experience and training, such a change in activities too often leads to ineffective habits that are damaging to both the scientific career and the organization.

It is neither good nor bad to accept more responsibilities over time. Rather, we should be aware of the factors that ought to drive decisions concerning which and how many responsibilities to accept, and make a conscious choice about the direction in which we would like our career to grow.

14.5 GENDER ISSUES

We consider the material in this section to be equally important for all young scientists, whether men or women. Moreover, much of the material pertaining to gender inequality discussed here is relevant also for other under-represented groups in science, regardless of whether that under-representation is associated with race, social status, religion, nationality, or physical disability.

Breakthroughs in science have been made by researchers of both genders. Far before the age of woman's liberation, Marie Curie carried out her seminal work on radioactivity and the discovery of the elements radium and polonium (Fig. 14.2). Marie Curie is, to date, the only recipient of *two* Nobel Prizes – in two different fields: physics and chemistry.

Despite the great scientific contribution of female scientists, a gender inequality exists in the development of scientific careers. This is illustrated in Fig. 14.3, which shows the percentage of male and female scientists as a function of academic position in the European Community for the years 1988–1999. As shown on the left side of that figure, the number of female undergraduate students actually exceeds

FIG. 14.2. Commemorative stamp in honor of Marie and Pierre Curie for their discovery of radium.

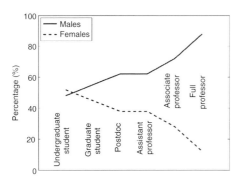

FIG. 14.3. The gender distribution for students and academic faculty in the European Community for the period 1988–1999 as a function of academic rank. (Source: Third European Report of Science & Technology, Indicators 2003, European Commission, Research DG.)

that of male students. Already, however, in graduate school the male students dominate, at least in numbers. This trend continues throughout the scientific career; of the full professors, only 12% are women! Since 1999, the gender participation has become more balanced, but this pattern of increase in the percentage of male faculty with academic rank persists. Clearly, many women must be leaving academic positions over time. Because of the shape of the curves in Fig. 14.3, this pattern has been called the *gender scissor*. There is no single reason for the imbalance in gender participation in science, but we list a number of factors that contribute.

- *Hidden expectations.* In the past, the default assumption of many was that the scientist is a man, and this assumption remains pervasive. We give two examples of the preferential recognition of the male scientist. In his quote in Section 14.3, Thomas Kuhn speaks of "men who achieve fundamental inventions of a new paradigm." Why not say "women who achieve fundamental inventions of a new paradigm" instead? As a second example, note that the stamp in Fig. 14.2 mentions "Pierre and Marie Curie" instead of "Marie and Pierre Curie." In the absence of objective reasons for choice of order, such as that in referring to "men and women" as opposed to "women and men," one choice or the other, of course, has to be made. The point here, however, is that, in practice, the scientist often is assumed to be a male. Superficially, one might think this only a trivial linguistic bias, but in reality the situation is not so simple. As we argue in Section 5.1, words play an essential role in our creative power, and, by repeating a gender bias in the words used, we support a skewed view of the contribution of the different genders in science. It is for this reason that, in this book, we preferentially use "she" instead of "he" when describing scientists. Here is a question for the reader: did you find it puzzling, or even disturbing, when encountering "she" instead of "he" for the first time while reading this book? Also, were you on the lookout for whether the reference to "her" rather than "him" pertained to a negative quality (or vice versa)?

- *Competitiveness of the academic career.* As discussed in Section 14.1, the scientific career is competitive, and, as sketched in Fig. 14.1, the total number of scientists decreases as a function of rank. In the eyes of many, in order to rise through the ranks an *aggressive* approach is needed. It is interesting to consider the dual values we give to the word *aggressive*. On the one hand it has a negative connotation because of its association with violence, but on the other hand, we often also consider it to be positive as it reflects a fighting spirit aimed at reaching goals. Because women often have more balanced personalities and suffer less from inflated ego,[11] female scientists might be less *aggressive* in the hunt for tenure and promotion than are their male colleagues.[12] The result of this laudable attitude, however, might well be that their male peers make more rapid career advancements.

---

[11] This is a reflection of the authors' bias.

[12] When, in contrast, the woman scientist is the one more solidly in pursuit of tenure and promotion, this is often interpreted as "pushy" behavior.

- *Traditional career role models remain.* The traditional role model in society is that the female takes care of children and domestic work, while the male goes out to generate income. Over the years, this traditional pattern has changed somewhat, but in practice even nowadays, women carry a larger part of child care than do their male partners. This, of course, leaves less time for female researchers to focus on their career, which further deepens the pattern of a male-dominated scientific community.

- *Gender bias in the search process.* Search committees tend to be dominated by men for the simple reason that there are more male than female senior faculty members. As an unfortunate result, there may be a bias to make a job offer preferentially to applicants who have followed the path of the traditional male career. Many academic search committees consider the number of publications an important factor in making hiring decisions. This could further disadvantage female applicants. For example, maternity leave, which, for biological reasons, is used more by mothers than by fathers, reduces the time spent on research and writing papers, and therefore reduces the length of publication lists. When evaluating candidates, some search committees attempt to take such personal circumstances into account in a meaningful and fair-minded way, while other committees might not bother to do so.

- *The two-body problem.* Since research is a relatively niche occupation, typically there are few openings in any area of research, particularly within a specific geographic region. For couples who both work in research, this creates the so-called *two-body problem* of having to find two challenging and satisfying positions within reasonable distance of each other. This is a severe constraint on the availability of attractive jobs, and the inability to find a job for the spouse, particularly when this entails a move to a new geographic area, can be a show-stopper in the hiring process. Since in our current society the woman is more likely to follow the man than the other way around, this poses a limitation on the type and quality of the jobs accepted by female researchers, with negative long-term consequences for their scientific career.

- *Lack of role models.* Our behavior is to a certain extent shaped by role models. It is common, for example, that students are fascinated by a research field through having been inspired by a particularly enthusiastic and charismatic professor. This is most likely to happen when the student can identify with that professor. Gender is among the many different factors that influence, even subtly, the degree of identification. Although this might seem incidental, female students and junior scientists benefit greatly from having role models of their gender. Such women provide examples of career paths to which to aspire, and they can have an especially large impact in mentoring young female scientists. Conversely, the absence of female role models can have the opposite effect of discouraging some women from pursuing a career in science.

This list of factors that contribute to gender inequality could be even larger when more speculative ones are taken into account.[13] A further problem is that some of the factors in this list enhance others. For example, the lack of female role models contributes to an under-representation of female scientists, which enhances the traditional male career model, which leads again to fewer female role models. These processes conspire to bring about what has been called the *glass ceiling*. This phrase captures the phenomenon that, while promotion is possible in principle, in practice many barriers to its realization exist. These barriers are typically not explicit, and only relatively few people intentionally create them, but they can and do hinder the promotion and tenure of under-represented groups.

So how do we break out of this pattern? We see two things that can be done. The first is to understand and create an awareness of the processes that are relevant; this is exactly the goal of this section.

---

[13] While writing this chapter we were surprised, and shocked, to see how easy it was to think of factors that contribute to a lack of diversity in the research community.

Second, it is essential to continue the push to break down the barriers that prevent the integration of under-represented groups in the scientific community. Progress in this area will take a willingness to depart from the traditional male career model and take active steps to promote the participation of under-represented groups. This raises a chicken-and-egg dilemma though. For whatever reasons, relatively few women and minorities have chosen to pursue certain scientific disciplines. Therefore, they will often be only a small minority of those applying for some academic position. A primary goal of a search should be to select and hire the applicant with the strongest credentials. For that goal to be satisfied, however, the odds when filling that position might not favor enhancing needed diversity in a department. To achieve diversity therefore would require that "all things not be equal." This would lead to a familiar dilemma of trying to balance two laudable goals – those of enhancing diversity and of striving to hire the most talented person for the position and thereby maintain the highest standard for the science pursued. Fortunately, these goals are not necessarily mutually exclusive, but in practice it can be a difficult balancing act to create effective and fair steps to increase the participation of under-represented groups. We must seriously work at such balancing.

14.6 CLOSING WORDS

The reader might get a feeling after reading this chapter that a career in research is a struggle and a continuous uphill battle. Such a career is indeed competitive, but there are two things to keep in mind. First, the rewards of a career in research are great. Discovering new things and creating an orderly explanation out of a confused pattern of observations is exciting. Moreover, a career in research can provide a stable and rewarding job. For those with a passion for teaching, academic positions offer a unique opportunity to combine teaching and research.[14] The authors both have a passion for teaching and get much

---

[14] Even in industry, teaching opportunities exist, through participation in in-house training programs and professional society short courses, and even through adjunct faculty positions.

satisfaction from helping young people better understand the world around them and prepare for a career in science or engineering. The joy of teaching is articulated by the following words of music teacher Linda Combellick.

> *A teacher is much more than a title or even a position. It is a way of being with others, a way of thinking about helping in this world. We facilitate, we begin the questioning or take it forward, we watch students grow right under our noses. The knowledge itself is secondary to the process they are moving through. It's about the whole being, not just about acquiring knowledge. THAT's a teacher from the inside out, a teacher from the heart ... To teach is to grow. We take in more and more of life, understand more and more and seek to understand more. We grow, we pass it on, we lend a hand to another's growth or just act as companion through their process.*

Many aspects of this quote are applicable to academic advising, as well, and express why advising is such a rewarding part of the academic career.

It is true that a career in research is competitive. It, however, need not be and should not be fiercely so. The overarching goal in research should not be to out-perform other researchers in the field. Rather, it should be to conquer the scientific problem that you've chosen to confront and to do that by being the best scientist you can be.

No job is perfect; some sources of frustration can undermine job satisfaction. Presuming that the number of complications and frustrations we encounter is not unduly large, the key issue is that of how we choose to deal with the unavoidable challenges and obstructions that we meet. Having a mindset that is effective for dealing with these issues while keeping a focus on the satisfying aspects of the job is extremely helpful. It could well happen, of course, that the problems encountered overwhelm the rewards of the job. In that event, the time might have come to move on and seek another position elsewhere.

It helps in the evolution of one's career to be flexible. We indicated in Section 14.3 that it can be advantageous to develop a diverse background by switching fields. For similar reasons it can be advantageous for career development to diversify skills. Taking training in management or pedagogy can be useful for researchers in both industry and academia. Such diversification not only helps create new opportunities, it can also lead to the discovery of interests that go beyond pure research. *Don't shy away from growth and exploration of new opportunities!*

15 Applying for a job

Whatever your current stage in graduate or even undergraduate study, like it or not, sooner or later you will have to work for a living – as if you haven't been working throughout your graduate school career. So, at some point the moment arrives to apply for a job. Much of what is covered in this chapter is the nuts and bolts of the job-application process aimed at a successful outcome of that process – securing a position that is right for you. But, more than just a tutoring of what to do and what not to do in applying for a job, the chapter aims to share our thoughts on subtle matters you might not otherwise think of but can expect to encounter along the way. Much of how to proceed and what to expect in applying for a job is common to careers in both academia and industry, but we shall also highlight differences.

As indicated in Chapter 14, the first job in an academic environment is likely to be a temporary position as a postdoc. Even then, soon enough the task of applying for a job will appear on the horizon, this time perhaps for a tenure-track faculty position. In industry, life-long employment with a single company used to be common, but this has changed considerably in the ever-more-dynamic world of today. The increased mobility of society applies largely to the job market as well; it is now common to switch employers periodically during a career, and each of these changes involves a new job application.

So, what defines a successful outcome from the job-application process? One aspect is obvious: a successful application leads to an attractive job offer. This, however, is not the whole story. At base, the application process aims at creating a match that is good – more than that, is *right* – for both employer and employee. Moreover, given your years of graduate study in preparation for a professional career in science or technology, you would hope that your first full-time position is a step toward satisfaction and accomplishment in that career.

A critical element for success in your job search is that you gain an adequate level of information about and insight into the position and the employer. You will want to understand the employer's expectations of you both in the short term and over time, the work environment, and opportunities for professional and career growth. The nuts and bolts of enhancing the chance for a successful job search include the writing of an effective application letter, preparation of a concise and informative resumé, a positive and instructive job interview, and, once a job offer is made, successful negotiation.

15.1 BE INFORMED

Being well informed about the job and the prospective employer should precede your decision on whether or not to apply, your writing of the application letter, your preparation for an interview visit, and your efforts at negotiation. Lacking such information, much of your time can be wasted, the application process has a reduced chance of success, and, worst, it could lead to landing in an undesirable work situation that ultimately leads to yet another job application.

Before deciding to apply for a position, strange as it might sound, first consider if you really would like the job. Applying for a job needs to be done with care; it requires much effort and can be an emotional drain. It therefore makes sense to apply only for positions you really want.[1] Here are questions you might pose to yourself. What does the job actually involve? What is its scope, and what are the responsibilities? What support, facilities, and authority do you receive to help carry out these responsibilities effectively? How much freedom and room for initiative and growth does the job offer?

Apart from these job-oriented questions, you want to be informed about the organization that might offer the job. Ideally, the

---

[1] We tacitly assume that you are not in survival mode, desperate for just any job. Underlying this assumption is yet another one – that you have chosen in favor of a path in science based on a desire to have a fulfilling job in discovery. Some scientists do find themselves in desperate need of a job; much of the advice in this chapter is applicable to that situation as well.

institution would be one that is held in high regard both broadly and in your scientific or technologic field, in particular. For a position in industry, you would also want to know the company's financial health and its future prospects. Much more than these dry basics about the organization, what is the essential *culture* of the organization? Yes, companies do have their individual culture, often set by the CEO and the company's tradition. Does the organization (whether industry or academic), or the group in which you might be working, have a friendly, professional, perhaps high-energy atmosphere, one that encourages and supports creativity? How does the organization treat its employees? Are there realistic prospects for growth and promotion within the organization? How hierarchical is the employer? Specifically, is it likely that you will be constrained and dictated to by superiors, or is there room for you to guide the way that your job evolves? How much committee work is expected, and how effective are these committees in influencing the policy of the organization? What salary and fringe benefits are being offered?[2] Perhaps high on your list is the location of the job or opportunities for travel or assignment elsewhere. Does it offer opportunities for the types of recreational and cultural activities that appeal to you? Most important, again, does this appear to be the job you really want?

The questions listed above are relevant for any job. There are others that are specific to positions in industry, academia, and national laboratories. For a job in industry the following issues merit consideration: What is the typical career path in the organization? How long do employees tend to stay with the employer, and why do they leave? How much room is there for training and growth? What is the employer's policy on publishing papers, taking short courses, and attending conferences? (Are these merely tolerated or actively encouraged?) How much freedom is there to choose the research topic and to

---

[2] Fringe benefits to take into consideration include health insurance, pension plan, and vacation time.

give direction to the research? Does the employer regularly transfer employees to other locations? If so, how frequently does this happen and to what sorts of location? To what extent would the organization's non-disclosure agreements limit your freedom to discuss your work with others? Are there restrictions imposed that could further limit career choices after leaving the employer?[3]

Issues that are more specific for academic positions include the following: What is the mix of research, teaching, and service that is expected? To what extent is there flexibility in the area of research? What sort of courses will you be expected to teach, and at what level? Is the position funded by the university, or does it depend on external funding? If the latter, what percentage of the salary depends on soft money? What percentage of your time can be given to outside consulting? For how many months per year is the appointment? (You might be able to seek further funding for activities over the summer, or be free to do consulting or simply relax and enjoy life during that time period.) Is the position temporary; that is, is it tenure-track, or tenured? How much research funding are you expected to generate? For a tenure-track position, what is the policy for making decisions on tenure and what fraction of tenure-track faculty actually gets tenure? What are the policies for taking sabbatical leave?[4] For example, how often can sabbaticals be taken, for what duration, and how much of your salary can be covered by your home institution while you are on sabbatical?

The above lists of questions are perhaps bewildering, yet you will want to be informed about these issues. It is of course difficult to have all questions answered before composing the job application, and, indeed, a goal of the full application process should be to answer

---

[3] It might seem surprising that an employer can legally control the career opportunities of its former employees, but as will be discussed in Section 15.5 this does happen and is something to be aware of.

[4] Many universities allow their faculty to take a leave about every seven years. This leave is aptly called a *sabbatical leave* and has the goal of starting up a new direction in research and/or teaching.

these and other questions. The job interview itself, discussions with the organization's employees during the interview, and a meeting with a human resource officer offer opportunities during the application process to address these questions. It is, however, often possible to obtain much information before writing the application letter. The internet is a rich source of information, but it is especially useful to talk with current employees of the organization, with its former employees, and with colleagues in other organizations in the same field.[5] Make sure to cross-check information when you have reason to believe that it might be biased either favorably or unfavorably. These conversations often give information that cannot be gleaned from websites and glossy brochures, and can be particularly reliable and pertinent to choices and decisions you will be facing as a potential employee.

With all these considerations, the decision process might be a difficult and frustrating one, but how pleasant it is to be in the position of having to face such choices.

15.2 THE APPLICATION LETTER AND RESUMÉ

Formal application for a position starts with an application letter, accompanied by a resumé or, for positions in academia and national laboratories, the larger, more complete curriculum vitae (CV).[6] All these documents need to be prepared with great care. Search committees and human resource officers typically are confronted with large numbers of applicants, and of necessity their initial decision to invite applicants for interviews is based on just what they see in the application letters, resumés, CVs, and possibly letters of reference.[7] If the

---

[5] Your professors could well be quite knowledgeable about characteristics of companies that specialize in technical fields aligned with your interests, and with universities with programs that are similarly aligned.

[6] The term *curriculum vitae* stands for "the program of your life."

[7] You might find yourself in the fortunate position of having already been sought-after by a potential employer for your recognized expertise in your field. Nevertheless,

application letter and resumé are not sufficiently strong, the candidate is unlikely to be invited for an interview, ending the application process for the position.[8]

15.2.1 The application letter

An effective application letter meets the following criteria:

- *The letter must look professional.* Most often, a search committee or company's hiring department initially sees only application letters and resumés; this material therefore conveys an essential first impression that you want to be positive. The letter and resumé must be well written (specifically, they should be clear and concise) in a font that can easily be read. They must be pleasing to the eye and convey a sense of professional expertise. A letterhead, when appropriate, and a simple layout contribute to creating a professional impression. *Run a spell-check before finalizing the letter.* Besides being unprofessional, typos suggest an indifference to the application effort. Be aware that the application letter and resumé usually will be photocopied. Therefore, using, for example, dark-brown letters on tan-colored paper is likely to lead to poorly readable copies. Make sure that the material you provide also looks professional after being copied, and don't rely on color to convey essential information because photocopies are likely to lack color.

- *Explain why you are the perfect candidate.* The search committee is seeking the perfect candidate from a large pool of applications. You should therefore ask yourself what could make you a top candidate for the job to which you are applying, and convey in your letter why you are particularly well suited for the job. In doing so, however, avoid

even when your application has been invited, you will probably still need to provide a well-crafted and informative CV or resumé.

[8] It is not uncommon for the employer to receive several tens, or even hundreds, of applications for a single position. Search committees face the difficult task of having to select only a few applicants from this large pool to invite for interview visits. For this reason, it is essential that your application letter and resumé stand out in this pool.

using statements that make explicit claims about your personality traits, such as "I am hardworking and reliable" and "I am trustworthy." It is much better that your letter conveys your accomplishments and traits in such a way that they are implicitly evident to the reader.

- *Show passion and a vision.* There is an immense difference between "just doing the job" and carrying it out with a passion and a vision. The latter takes the work to a higher level in terms of usefulness for the organization and value of your contribution to yourself and others. The applicant with passion and a vision comes across more convincingly than does one who conveys a sense of dull submission to the work that needs to be carried out. When applying for an academic position, for example, it is crucial that you convey a vision for the research that you hope to do, and a passion for teaching when that is part of the job description.

- *Display a genuine interest in the organization.* The employer seeks an employee with not only the right technical qualifications, but also a genuine interest in and engagement with the organization. Therefore, in addition to focusing on your individual interests as a potential employee, the application letter should also convey knowledge of the larger goals of the organization and ways in which you can contribute to these goals, and convey what it is about the organization that has drawn you to apply.

- *Be to the point.* Again, search committees and hiring teams are staffed by busy people who receive many applications and therefore need to move through the search process efficiently. Application letters that are verbose and long-winded create an unfavorable impression. Some applicants make the mistake of listing in great detail all of their achievements, no matter how incidental. Apart from boring those poor souls who are attempting to read the letter, this can create an unfavorable narcissistic impression.

- *Don't be too modest.* Just the same, if you are the right person for the job, you need to convey this through your list of skills, qualifications, and achievements. False modesty is ineffective for bringing these points across. The job application is not the moment to be overly modest, so don't hesitate to toot your horn!

Some of the criteria listed above are conflicting, and attaining the balance needed in a quality application letter is difficult. Writing an application letter takes some research to learn about the employer, and requires deliberate thought about how you best fit into the position. As with any writing, it is helpful to put the letter away for a while after you have written it. It can be particularly helpful to ask colleagues to read, proofread, and give you feedback about the letter. This can be especially so when a person you ask is employed by the organization to which you are applying, provided that this constitutes no conflict of interest for that individual.

15.2.2 *The resumé*

The resumé is a document that provides factual information about your background. Just as for the application letter it must look professional (also after being copied), convey the necessary information, and be succinct. Out of consideration for the patience of search committees, who usually must review large numbers of applications, the resumé should not be unnecessarily long and should contain only the relevant information.[9] A template outlining the basic elements in a resumé is shown in Fig. 15.1. There is no fixed format for a curriculum vitae, and many templates can be found through the internet. We use the template of Fig. 15.1 to discuss the essential elements of both resumé and CV. Resumés are usually sufficient when applying for positions in industry, whereas CVs are almost universally required for positions in academia and at national laboratories. The essential difference between a resumé and a CV is that of length and detail of material contained. Whereas the resumé should be concise,

---

[9] Also to aid in readability, take care in choice of spacing, indentation, and use of bold font for headings. These aids to the esthetics help the reader quickly identify the main points in the resumé.

CURRICULUM VITAE

Name
address

Telephone:
Fax:
Electronic mail:
website

Education:

-

-

Positions held:

-

-

Professional honors:

-

-

Professional activities:

-

-

Publications: see attached list containing N internationally refereed publications and M other publications.

Grants: see attached list of K externally funded projects

References:

-

-

FIG. 15.1. A template for a curriculum vitae or resumé. The resumé would contain less detail, in particular about publications and grants. In practice, the CV has a length of a few pages.

typically no more than one or two pages, the CV can be extensive, listing education and all relevant positions (with dates shown), honors, professional and academic activities, short courses taken, research conducted, publications, oral presentations at professional society meetings, invited lectures, grants, courses taught, service to professional societies, and committee participation.

First and foremost, the resumé and CV should contain the applicant's name and contact information. Don't forget to include that! In some countries, such as Germany, it is common to place a small photo on the resumé as well. One can add other information such as age, gender, race, and marital status as well. But don't do this in the USA! Because of anti-discrimination laws, employers usually are prohibited from taking such information into account, and, in countries where lawsuits are commonplace, they prefer not to know such personal data. Don't hesitate, however, to add personal information to your resumé when such information is relevant to the target job or for doing justice to who you are.

Your education conveys your background and training. In general, it suffices to limit this list to education after leaving high school. Education is not necessarily restricted to academic degrees; for example, professional training received in industry could be an essential part of your education as well – if it's relevant.

The list of previously held positions is of interest to the potential employer. When you have held positions in academic or industrial research, it is easy to compile this list, but, again, in the resumé avoid unnecessary descriptive detail, and restrict the list to those positions that capture the core of your professional experience. For those who have just completed graduate school, the list of relevant positions held is often limited. Don't forget, though, to mention research or teaching assistantships and summer jobs that you have had, provided again that they are pertinent.

In the scientific community, honor and appreciation is bestowed on individuals in many different ways. Universities, companies, and professional organization have prizes, awards, and fellowships. Students organizations often select a "teacher of the year" or have other ways to express appreciation for a particularly gifted and devoted teacher. Funding organizations usually issue prestigious fellowships and grants, and scientific journals sometimes highlight papers of particular scientific merit. Since these are indicators of professional qualities and of the high regard that others have for your contributions, it is appropriate and advantageous to list such honors received on the resumé.

Apart from official positions held, scientists often contribute to their profession in other ways as well. Such activities include editorships, serving on advisory committees, the organization of scientific meetings, teaching at prestigious summer schools, mentoring of under-represented groups, and outreach to the general public or to primary or secondary education. Participation in these activities not only indicates professional expertise, but also points to an engagement that goes beyond the level of just doing a job and suggests value perceived by others. The CV allows more space than does the resumé to elaborate on honors and awards, as well as on professional activities.

We indicated earlier the importance of publications in developing a career in science. Of similar merit is the track record for attracting funding for research and teaching. Since a resumé has to be concise, it is not the place to list publications or grants received. The list of publications and funding generated is appropriately placed in the curriculum vitae, perhaps in an appendix. A reference such as "attached is a list with 20 internationally reviewed publications and four other publications" can capture your scientific productivity in a single sentence of the resumé. The same can be done for the funding you have been able to attract, e.g., by citing the number of funded projects or the average annual funding received.

Potential employers usually will wish to contact professionals who know job applicants well. References from individuals known to the employers are particularly valuable. For the list of names and contact information of colleagues who are willing to be contacted as a reference by a potential employer, choose people who know you and your work well and who hold you in high regard. Always include in the list only the names of individuals from whom you have previously asked for, and received permission from them, to act as a reference. First, it is a simple matter of courtesy to ask if they are willing to do so; writing a letter of reference can take considerable time and effort. Second, it could happen that a potential reference has some reservation about your skills or qualifications. By asking people if they are willing to serve as a reference, you give them the opportunity to politely refuse to do so if they wish. In practice they might do this by offering some excuse for refusing, for example by telling you they are terribly busy or don't feel they know you well enough.

Consider what is absent from the skeletal resumé in Fig. 15.1; hobbies, for example, are not listed. Potential employers could well be interested in learning personal information about you, such as that you like hiking, cooking, and dancing, but that information is super-fluous – indeed clutter – on the resumé or CV, and can be conveyed later, during the interview visit. Having said that, there could be per-sonal activities that indicate skills or qualifications that are of true relevance to the job; you should have no reservations about listing those. Also absent from the template in Fig. 15.1 are trivial job skills. It is unnecessary, and in fact can create a negative impression, for you to inform a potential employer that you know how to titrate, have experience with Microsoft Office, or know how to program with Mat-Lab. There could, of course, be professional skills that make you stand out among your peers, and it could be valuable to list those, but list-ing standard job skills clutters the resumé and creates an unfavorable impression of making a big deal over standard skills. The same holds for languages that you may have mastered, unless these languages are an essential part of the position for which you are applying, for example if you will work abroad.

Despite your best efforts at writing an application letter and resumé, you might not be invited for a job interview. Often, the sim-ple reason is that there are so many applicants and only a few could be invited for an interview. Even so, it can be helpful for improving future job applications to learn why preference was given to others. Do not hesitate to contact somebody on the search committee for the job for which you had applied to ask for feedback on your application. Was the application clear? Did it convey a professional impression? Were cer-tain key components missing? Was there something in the application that was off-putting? Are there any suggestions for improvement? Don't hesitate to seek such potentially valuable feedback.

15.3 THE INTERVIEW VISIT

All the time, effort, care, and creativity that you've put into the "paper stage" of the application process – writing and submitting your application letter, with enclosed resumé or CV – is aimed at just opening the door into the theater, the hoped-for interview visit. Some potential employers choose to include an intermediate stage, a telephone interview, prior to the main event – a visit to the institution's office for a day or more of face-to-face meetings between you and employees. This is the all-important opportunity for you and the organization to emerge from paper to gain a more complete sense of one another. It can be expected that the organization will be inviting a number of candidates for interviews for the position, so there is no stage in the application process more crucial to conveying a favorable impression than during the interview visit. Creating such an impression, however, is not the only goal of the interview visit; it also serves as a source of information about what the job exactly entails and about the work environment provided by the organization.

So, the utmost care is necessary in preparing for the interview. The visit for a job application typically has a duration of 1–2 days and consists of one or more seminars, an interview with the search committee, several one-on-one conversations with faculty members and administrators, dinner with interested parties, and, for academic positions, one or more meetings with students. Make sure you know, as clearly as possible in advance, what to expect during your visit. Who will you meet?[10] What are the background and interests of those you will meet? Will you be expected to give a seminar? If so, what are

---

[10] As a rather green student, one of us attended a prestigious dinner in his honor at the Dutch Science Foundation and asked his neighbor at the table "do you work here?" The sobering reply was "I am the President of the science foundation." This is not the way to make a favorable impression.

the scope, audience, and duration of the seminar?[11] Will you meet an officer of the human resources department, and what information do you need to know from that department? In practice, the invitation for a job interview will be issued by the chair of the search committee or a human resources officer, and it is perfectly valid to ask at that time, before arriving for the interview, what you can expect and what is expected of you during the interview. Some more extensive background research on the organization than you had needed before the invitation will be helpful preparation for the interview visit. Your knowledge about the organization will help in conveying that you have taken the care to become thoughtfully informed.

Because we convey much information and personal impressions through our body language and appearance, it will do you well to pay attention to these. Make sure to be appropriately dressed. This does not necessarily mean to be overly formally dressed, but be aware of what is likely to be expected. When in doubt, it is better to err toward the more formal rather than the informal. The simplest of considerations regarding body language can help with the impression you hope to convey. Offer a firm handshake without crushing the hand of your host, look people in the eye, and make sure that, during an interview meeting with several people, you are seated in a location where you can see everybody else. These seemingly inconsequential measures not only help with all-important first impressions, they can help keep you in a positive, relaxed frame of mind.[12]

It is beneficial to be active during the interview and to drive the discussion. This helps in gathering the information that you need for making further decisions and conveys a keen sense of interest in the potential job. Through your research in advance of the interview, compile a list of questions, and don't hesitate to ask all of your questions.

---

[11] As with any presentation that you give, know your audience for the seminar, and tailor the content and tone of your talk to that audience.

[12] Important as the job interview is, it is not the sole opportunity of a lifetime, nor will it lead to the decision of a lifetime; other such opportunities will come along. So, relax and enjoy the interview. (But don't be cocky about it.)

Keep in mind that the interview is a two-way process; you are evaluating the organization and the work environment it likely offers as much as individuals in the organization are evaluating your potential as an employee.

Few scientists are hired for being passive and subdued. For this reason it is important to make a lively and enthusiastic impression. Be outgoing during your visit, don't hesitate to ask lots of questions and engage those you meet in lively conversation. It is essential throughout, however, to stay true to yourself. Pretending to be different from who you really are will be unconvincing. Otherwise, in the event that you nevertheless are offered and accept a position with the organization, you might well be hired for the wrong reasons. This can lead to a poor match between you, your job, and the employer, a recipe for disaster over the long term. Therefore, don't build a facade, pretending to be somebody different from who you really are.

15.4 NEGOTIATE!

The job-application process, one hopes, results in an offer. You are, of course, free to accept or decline such an offer. In practice, before accepting one, a negotiation should take place, with the goal of refining the offer into one that is acceptable to both employer and employee. We want to stress the importance of carefully negotiating before accepting a job offer because the terms of employment are essentially fixed once a job is accepted. Of course, there will be occasional raises and promotions, but the power to negotiate is small once you have started on a job. For this reason, consider a job offer as the primary window of opportunity for obtaining favorable terms of employment. The next opportunity might arise only when switching jobs, something that is not necessarily in your best interest.

So what is there to negotiate? For both academic and industrial jobs there are the main terms of appointment. The job offer will likely come with fringe benefits such as health insurance, number of vacation weeks per year, a pension plan, and an employer-supported savings plan. In general, it is difficult to negotiate these benefits since they usually are the same for employees within a given organization. Of the terms of employment, salary is clearly a major one and can, within limits, be negotiated. It is good to have an idea of the salary typically offered in the marketplace for the type and level of job being offered, as well as the salary that your employer offers for comparable positions. Knowing these numbers sets a realistic range for the salary to aim for. Quite often, companies offer larger salaries to new hires than those that they pay their faithful and productive long-time employees. Because of this unfortunate practice, the salary "increase" that you negotiate at the time you are hired or change employment could well be the largest percentage increase that you will receive during your subsequent time of employment.[13]

Financial compensation is but one aspect of a job worth negotiating, particularly for positions in academia. Another might be the content of the job, for example, the amount of teaching required in an academic position. Others, again in academia, include facilities, equipment, and support staff necessary in order to conduct research most effectively. When the job involves setting up a laboratory, one needs space for the facility. The time when an offer is negotiated provides the best opportunity to ensure that the required facilities are available. The same holds for needed support staff, such as an administrative assistant and laboratory technicians. Make sure to get the facilities, equipment, and support you need in order to carry out your job effectively.

Also, for academic positions it is common to receive *start-up funds* and *discretionary funds*. Start-up funding can be used to build up the research. Common uses for such funding include building and

---

[13] Likely, the following advice won't apply for those motivated by factors other than salary. Nevertheless, we caution you to avoid the temptation to give too much weight in the choice among alternative job offers to the one with the highest starting salary. The starting salary is just that, *starting*. Your ability to command a larger salary later in your career depends on other factors, primarily on the value you bring to the organization, value that could well derive from your love of the work that you do and from the environment that the organization provides for you to thrive in that work.

equipping a laboratory, purchasing computers, hiring of graduate students, assistants and technicians, and covering part of the summer salary for the first few years. Discretionary funds are allocated yearly and can be used, for example, to cover the costs of doing the research, hiring administrative or technical support, and attending conferences. Some universities offer generous discretionary funding, while others offer virtually none – or provide it only when teased out through negotiation. Whatever amount you agree on through the negotiations, don't forget to put in a clause to have the funds corrected annually for inflation.

Many researchers, especially those in academia, are uncomfortable with negotiation. This might be due in part to a sense that we do research as a fulfillment of our inquisitive nature, which does not necessarily need to be awarded with generous terms of appointment. Indeed, for many researchers the primary satisfaction of the job is not the financial compensation. This does not mean, however, that one should pay little attention to the terms of appointment or to negotiating for the best possible initial conditions; negotiation is neither inappropriate nor immoral.

Many scientists don't have much experience with negotiation. It can be worthwhile to read a good book on this topic before applying for a job. Practice by playing Monopoly with friends and bargain hard; and, if you want a tougher exercise, haggle with car dealers about the price of a used car.[14] Your negotiating position is strongest when you have other job offers and the potential employer knows that he or she is competing with others. Paradoxically, having other job offers can make you appear more attractive in the eye of prospective employers, with the enhanced prospect for a job offer and the quality of that offer. Therefore, when in the position to do so, don't hesitate to talk to several potential employers. We recommend that you conduct your negotiations in an amicable and straightforward manner, with the understanding that both you and your potential employer have a

---

[14] Of course, you don't have to actually buy the car.

common goal – to obtain terms that are conducive for you to perform at your best in your new position.

15.5 BEFORE SIGNING A CONTRACT IN INDUSTRY

As any new employee certainly expects, the first day on the job will include necessary paperwork. There will be forms to complete with the human resources department – forms for the various insurances, to get onto the payroll, and to enroll for other benefits. Some, not all, companies will also have you (the new employee) sign an agreement covering issues of patent ownership and confidentiality. Such an agreement might cover any of a number of issues of importance to the company. An example is an agreement that makes clear that the company is the owner of any idea in your technical field that you develop while an employee. Another might require that you do not divulge or disclose to others outside the company any information that is proprietary to the company, either while you are employed there or afterward, if you have changed jobs and work for a new employer. Further, you may not use such confidential information outside the company as long as it is confidential. Such agreements, which must be signed upon starting a new job, are commonplace in industry. Their purpose is to cover the legitimate right of the company to protect its proprietary ideas and technology.

But there are other, considerably more demanding, requirements that some companies might attempt to impose, ones to be especially aware of and to take into serious consideration in the decision process. One such part of the agreement might stipulate that you (again, the new employee) agree that, for some stipulated time period after leaving the company to take employment elsewhere, you will not work in areas of research or technology in which you had been directly involved while working for your present company. A more extreme requirement that some companies attempt to impose is one that states that you are not allowed to work for, or consult for, a competitor for a specified period of time after leaving the company. In some such so-called "non-compete" agreements, the time periods for such requirements can be significant – as long as one and even two years. You should think very carefully before signing such a binding agreement – *and it can be binding.*[15]

Perhaps you would not be troubled by having to sign a binding agreement that precludes your working for a "competitor" for N years. That, of course, is your choice. The essential point here is that you be well aware of the binding nature and implications of agreements that have the potential to exercise control over your career options beyond the term of employment with your new employer. Then, the choice can indeed be *yours.*

The non-compete agreement that you sign when starting with a new employer (or, as could happen, after you've been with the company for some time) is written primarily to protect the company's interests – not yours. So you need *time* to review the agreement. Aside from this time requirement, *timing* can be even more important. Suppose you've just graduated from a university in one city and have made your choice to work for a company in another city. The company agrees to pay your moving costs. You move, and show up for your first day's work. At the human resources department, you are asked to sign the necessary paperwork to get started. It all seems routine, so your inclination might be to sign these presumed *standard* papers without giving them much thought or reading them with care. Among the papers is an agreement with a paragraph that precludes your working for a competitor for N years after leaving this company, if you should leave. Is this the moment when you would first wish to learn of such a binding agreement? Even if you should ask to take the agreement home for review, what will you do in case you find some of its terms unacceptable? You've already turned down other offers, moved to a new city, and have the moving van with your household goods on its way.

---

[15] Non-compete agreements can be successfully challenged in court, but at the cost of time, anxiety, and money.

So, here is the core advice: *Before* you make your decision among different job offers, and *before* you've left your university town or home town – indeed, as an essential component of your decision process – ask the companies that have made job offers to send you copies of any and all agreements that they will have you sign when you join the organization. You would then have the opportunity and time to seek legal advice in case you find a possibly troubling non-compete clause in the agreement of a company for which you otherwise are attracted to work.

The cautions raised here – about (1) items that might appear in a prospective employer's confidentiality agreement, and (2) the stage in your decision process when you would want to know the contents of such an agreement – offer no opinion on the legalities of such restrictive agreements. At the time you sign any agreement, however, you must assume that it will be binding.

Are such agreements ethically right and justifiable? The requirements written into confidentiality agreements range widely – from being quite mild and reasonably undemanding to being beyond the bounds of what we personally might think proper. Whatever the specifics, these agreements are written in the interest of giving the company the protection that it feels *it* needs.

Some companies require you to sign a binding non-compete agreement and others not, and some of the agreements have potentially severe implications for your future career, while others do not. The essential point here is, again, that it is much in your interest to be aware of the likelihood that some of your prospective employers will attempt to require your signature on a binding non-competitive agreement that can severely limit your choices for future employment if you should leave the company. That knowledge gives you the *opportunity* to request and study the content of the agreement in sufficient time to allow you, early on, to factor it into your employment decision.

16 Concluding remarks

The best time to take action towards a dream is yesterday; the worst is tomorrow; the best compromise is today.

Simon, 1998

Whether you are early in a Ph.D. program or further along in your graduate studies, an undergraduate contemplating graduate school and a career in science, a recently anointed Ph.D. embarking on a scientific career, or are somewhere beyond in mid-career, our hope is that various of the suggestions offered in this book can be of help toward your goal of a successful and satisfying professional career.

Much of the advice on doing research contained in this book involves practical skills. Regardless of the practicality of this advice, neither this nor any other book can provide a recipe that guarantees success. As argued in Chapter 2, despite its foundation in logic, science is driven by inspiration, insight, intuition, and creativity, all combined with technical expertise. No cookbook-style set of instructions based on a combination of just these skills, however, can offer a young scientist the guarantee that these ingredients, when mixed, will yield a fruitful and satisfying professional career. Our careers depend on not only our scientific talents, but also our personal ones and our attitude in life. In closing, we offer advice on the development of a mentality that helps foster success in the professional and personal aspects of a career in science, engineering, or humanities.

16.1 CREATE YOUR OWN LUCK

Luck is where preparation meets opportunity.

Seneca, 5 BCE – CE 65

Some people experience repeated success while, for others, life is much more difficult. What determines this difference? Are some people just lucky, and others not? In only a limited sense can our fortunes be attributed to luck. Our goals and aspirations for a successful career in science in general have two key components: the circumstances must be right, but more fundamentally we must create our own luck. Some people happen to be at a laboratory where everything is ready for breakthrough work; they might meet colleagues who give them important advice; or they could receive support from others that turns out to be crucial for carrying out the work. Such external circumstances are propitious for achieving success, but seldom is this enough. For some, the external circumstances seem ideal, so those individuals could well be considered lucky in the eyes of their peers, yet despite these conditions they are unable to realize the full promise of their situation. Circumstances are of little help if one doesn't put them to proper use.

More often than not, you can create the conditions that help you most. For example, suppose, in a neighboring department, a seminar is given that seems perhaps only vaguely related to your research. Do you go to that seminar or do you keep on working on your project because you are so "busy"? Without having any knowledge that it will happen in advance, possibly what you hear in the seminar, if you choose to attend it, could influence the course of your research. Your choice to attend could provide seeds of "good luck" for you and your research. During the seminar, you can ask the speaker questions, and afterward you might be able to talk with the speaker if you wish. Your attendance at the seminar and the availability of the speaker present you with an opportunity, but that opportunity can be of value only when you exercise it to advantage. Often in practice, unsatisfyingly, the seminar that you have chosen to attend turns out to be disappointing and indeed a diversion from your research. That is all part of the game. Any number of disappointing seminars, however, will be greatly offset by that one seminar that gave you a truly new insight or where you met somebody who later turned out to be of invaluable help in your work.[1]

---

[1] Here is an example of how a little initiative can pay off. One of us (Roel Snieder) was working on extracting the response of mechanical structures from random vibrations. In the absence of a data set, this work, while interesting in itself, seemed of little other value. A guest seminar speaker at a nearby institute showed recorded

The example of the seminar serves to illustrate the second, more important, component of luck; although external circumstances might be favorable, you must also create and exploit your own opportunities. Without this second component, your chances of turning out to have been considered lucky are minimal; therefore work to create your own luck. Opportunities come regularly in anyone's life, and often they are only briefly available. Being only opportunities, it is up to the individual to spot them and put them to use. Moreover, because they usually appear for only limited windows of time, proactive steps are needed in order to convert these opportunities into the favorable circumstances and good luck that you would like to have come your way. Such windows of opportunity come in many forms during the course of a scientific career. A visit to your university by an inspiring and influential scientist offers the opportunity to establish contact with that person. In a research project, a measurement might be puzzling and surprising, presenting a brief choice to either pursue this in greater depth and perhaps make a breakthrough, or ignore the anomalous data. An email message might arrive with the invitation to apply for an internship in industrial research. Do you ignore this opportunity or pursue the internship and possibly land a wonderful job?

People fail to take advantage of windows of opportunity for any of a variety of reasons. We might not realize at the moment that an opening exists; sometimes we are indecisive and let the moment go by; and fear can be another factor in allowing an opportunity to pass by. Being attuned to recognizing opportunities when they arise and having the courage to act decisively are key to converting opportunities into conditions that one might call "good luck."

16.2 LIFE IS A BOOMERANG

In Section 5.1 we discussed how thoughts lead to words, and words to actions. By repeating a thought, word, or certain action often enough

vibrations in a building. Talking with the speaker, requesting and later receiving the data, led to a study that produced two publications. Showing an early draft of one of these papers to a colleague in industry produced the invitation to submit a proposal, which subsequently led to a 3-year research contract supporting fruitful research. None of this would have happened without having attended the seminar.

it becomes internalized as part of our character. The character that we develop, moreover, defines our predominant mental state and can influence the development of our physical body as well. If you are permanently angry, it is likely that your face will grow into an angry one; in contrast, if your basic attitude is one of trust and enthusiasm, you are likely to develop a positive and energetic appearance, one that radiates a faith and optimism that things will work out well. The character and characteristics that we develop influence how we act and react to others, and contribute to fashioning how our lives evolve. In this sense, our character helps give shape to our destiny. We use the word *destiny* with some trepidation, as we don't want to suggest that the course of our life is pre-ordained. The chain of events initiated by our thoughts constitutes a path that helps give shape to our life, molding it toward the destiny that, one hopes, we desire. The chain of events given in Section 5.1 might thus be extended in the following way.

$$\text{Thoughts} \Rightarrow \text{Words} \Rightarrow \text{Actions} \Rightarrow \text{Character} \Rightarrow \text{Destiny}$$

This chain has everything to do with creating our own luck. When our thoughts are focused on creating new opportunities, learning new things, and meeting interesting people, the corresponding actions are steps toward good luck in life. This does not necessarily imply that we *will* succeed in all we want to do. When our actions are the result of positive thoughts, however, they inspire faith in ourself and those around us, thus becoming a driving force in the further development of our character and, ultimately, our destiny.

People around us tend to mirror our own attitudes. When we are short-tempered, others don't enjoy our presence and, in response, tend to react in a brusk manner. In contrast, an active interest in others draws to us people who are likely to be interested in us and in what we do, as well. By being inquisitive and enthusiastic in our daily work as scientists, over time we attract colleagues and students who share this attitude and thus contribute to the mutual success of our scientific endeavors. The words of actor Marcello Mastroianni[2] that "life is a

---

[2] From the movie *"Cos come come sei."*

boomerang" capture the truth that those around us often reflect our attitude, and that events that shape our future are, to a considerable extent, caused by our own actions and their underlying thoughts. By consistently maintaining a certain attitude, we encourage a similar frame of mind in those around us. When used in a constructive way, this offers a wonderfully rich opportunity to help develop our environment into one that is positive, creative, and nurturing. This is summed up in *leadership by example;* one of the most effective ways to give direction to a group, perhaps to the destiny of the group.

16.3 THINK OF — AND BE — A ROLE MODEL

That which we are, we shall teach, not voluntarily but involuntarily.

Emerson, 1841

As a simple exercise, think back to your days as an undergraduate student or perhaps to your time as a graduate student. Invariably, there will be a teacher or mentor who you particularly liked and admired. This person probably had a good measure of both charisma and empathy for students. The odds are high that, apart from that teacher's in-depth technical expertise, words such as *love, enthusiasm, trust,* and *commitment* come to mind from your recollections. This is no coincidence; these character traits are essential for a balanced and positive approach to our professional and personal lives. We have focused here on the professional aspects of these attitudes, but much is comparably applicable to our personal lives.

We argued in Chapter 3 for the great importance of choosing a research topic for which you have a passion, and for picking an adviser and other co-workers whom you like as individuals. We are at our best when we love what we do, and we enjoy working with colleagues whom we respect and, simply, like. The love for our work is essential to giving ourselves fully to it; it gives the involvement needed to fully tap into our creativity. Without love and care, what we do can become empty and lose direction, certainly not well-spring conditions for creativity. It is the love for our work, for our colleagues, and for those who benefit from our work that gives meaning to our work, well beyond the self-centered pursuit of promotion and career-opportunities. A former colleague of ours at the Colorado School of Mines, Alexander Kaufman – a passionate, devoted, and loved teacher (with no shortage of charisma) – mentioned to one of us that

> *You can teach well only when, while standing in front of a classroom and looking at your students, you realize that, without exception, you truly love each and everyone of them.*

His principles for what it takes to teach well – being fully focused on our students, recognizing their needs, and responding to these needs in the most constructive and positive way we can think of – hold comparably for research.

Research cannot be done well when it is carried out with the mentality of someone who dutifully completes the required hours at work. Without enthusiasm, one's actions tend to be lackluster and uncommitted. The enthusiasm of a researcher for her work is critical not only for the sense of well-being of the researcher, but also for the quality of research. Enthusiasm is a great driver that pushes us forward, providing a spark, momentum, and a passion for our actions. Often it is our enthusiasm that makes us put in the extra effort needed to move research forward and make us work with the intensity essential for exploiting our creativity to its fullest. Moreover, since others tend to mirror us, enthusiasm often is contagious. An enthusiastic and inspired team leader (or team member) can be the critical factor for sparking the creativity and involvement of students and co-workers. Enthusiasm thus not only helps propel our own work, it is an effective motivating factor for those around us.

16.4 TRUST AND COMMITMENT

> *Trust thyself; every heart vibrates to that iron string.*
> Emerson, 1841

One can focus on threats and limitations, or on opportunities and growth. An illustration of this is skiing among trees, an exciting activity but one that can be dangerous, as well, because of the risk of colliding into a tree. One of us received the advice *to focus not on the trees, but on the space between the trees instead.* This advice promotes a focus on the solutions and opportunities (the space between the trees) rather than on the problem (the trees themselves). This illustrates the issue of trust. When carrying out research, we can focus on the aspects of our work that might go wrong, or we can trust that opportunities for us to find solutions will present themselves. The shame with being driven by our uncertainties is that we often attract the very problems we wish to avoid, just as a skier who focuses on the trees actually attracts the possibility of crashing into a tree. Trust pushes us forward and energizes us to make things happen because it gives a focus on the solution. We don't want to imply that one should not be aware of complications and problems that exist and need to be overcome; simply, it is much more effective to deal with these issues from a constructive basis of trust that these problems can be overcome rather than from a paralyzing attitude of doubt. An additional value of trust is that it allows us to move beyond our comfort zone into new activities. This helps us grow in new directions and explore different aspects of our skills and personality.

Commitment is the key ingredient that translates our love and enthusiasm for our work into actions. Inertia, in contrast, can prevent us from initiating a new project; unless we are committed to starting the project, it might never get off the ground. Once having started a project, we could possibly discover that much of the day-to-day work of research is an uninspiring repetition of routine activities, or we encounter at some point stumbling blocks and barriers that need to be overcome. These are unavoidable aspects of many lines of research, but these downsides are balanced by the joy of discovery and a career that, in the main, is successful and meaningful. The commitment helps us stay focused on the goals of the work and remain enthusiastic when unavoidable drudgery or setbacks are encountered and need to be overcome. Without commitment it is easy to waiver and not follow through on a line of research at sufficient depth to discover what is there awaiting discovery.

But there is more to commitment. When we stick to our guns and follow through on our plans, serendipitous events that greatly help us in our research often come our way. This is articulated beautifully by the mountaineer Murray (1951).

> *Until one is committed, there is hesitancy, the chance to draw back, always ineffectiveness. Concerning all acts of initiative (and creation) there is one elementary truth, the ignorance of which kills countless ideas and splendid plans: the moment one definitely commits oneself, then Providence moves, too. All sorts of things occur to help one that would never have occurred. A whole stream of events issues from the decision, raising in one's favor all manner of unforeseen incidents and meetings, and a deep respect for one of Goethe's couplets: "Whatever you can do, or dream you can, begin it. Boldness has genius, power, and magic in it."*

So, again, we invite you to think back again to the inspiring professor or mentor whom you encountered in your past, and the extent to which his or her personality was anchored in love, enthusiasm, trust, and commitment. Then think of yourself. What do you want to achieve as a researcher? What is the life that you want to live? Who do you want to be as a researcher and human being? Perhaps a mentor valued by young, inquisitive, budding scientists? Answering these questions and giving your life shape according to your answers is the real adventure. Enjoy the adventure!

Appendix A   Further reading

There are many books on the market with advice for graduate students and other researchers. Some are listed here in various categories.

General advice for graduate students:

- Bloom, D.F., Karp, J.D., & Cohen, N. (1998). *The Ph.D. Process, A Student's Guide to Graduate School in the Sciences*. New York: Oxford University Press. This is the book for every graduate student in the physical sciences to read.
- Bolker, J. (1998). *Writing your Dissertation in Fifteen Minutes a Day*. New York: Henry Holt and Company LLC.
- Booth, W.C., Williams, J.M., & Coulomb, G.C. (2003). *The Craft of Research*, 2nd edn, Chicago: University of Chicago Press.
- Davis, G.B. & Parker, C.A. (1997). *Writing the Doctoral Dissertation*. New York: Barron's Educational Series Inc.
- Feibelman, P.J. (1993). *A Ph.D. is Not Enough! A Guide to Survival in Science*. Cambridge MA: Perseus Publishing.
- Medawar, P.B. (1979). *Advice to a Young Scientist*. Basic Books. http://www.basicbooks.com.
- Peters, R.L. (1997). *Getting What You Came For, The Smart Student's Guide to Earning a Master's or Ph.D.*, revised edn, New York: Farrar, Straus and Giroux.

The scientific method:

- Gauch, H.G., Jr. (2003). *Scientific Method in Practice*. Cambridge, UK: Cambridge University Press.
- Goldstein, I.F. & Goldstein, M. (1984). *The Experience of Science*. New York: Plenum Press.
- Gula, R.J. (2002). *Nonsense: A Handbook of Logical Fallacies*. Mount Jackson VA: Axios Press.
- Hatton, J. & Plouffe, P.B. (1997). *Science and its Ways of Knowing*. Upper Saddle River, New Jersey: Prentice-Hall.
- Martin, R.M. (1997). *Scientific Thinking*. Peterborough, Ontario: Broadview Press.

- Robinson, K. (2001). *Out of Our Minds: Learning to be Creative*. London: Capstone Publishing Limited & John Wiley and Sons.

Bioethics:

- Beauchamp, T.L. & Childress, J.F. (2001). *Principles of Biomedical Ethics*. Oxford, UK: Oxford University Press.
- Levine, C. (2005). *Taking Sides: Clashing Views on Controversial Bioethical Issues*. New York: McGraw-Hill.
- Mappes, T.A. & DeGrazia, D. (2000). *Biomedical Ethics*, New York: McGraw-Hill.
- Pence, G. (2004). *Classic Cases in Medical Ethics*. New York: McGraw-Hill.

Time management:

- Burka, J.B. & Yuen, L.M. (1983). *Procrastination, Why You Do It, What To Do About It*. Boston MA: Perseus Books.
- Covey, S.R. (1990). *The 7 Habits of Highly Effective People*. New York: Fireside Books. This is a must-read!
- St. James, E. (2000). *Simplify Your Work Life*. New York: Hyperion.

Communication:

- Alley, M. (2003). *The Craft of Scientific Presentations: Critical Steps to Succeed and Critical Errors to Avoid*. New York: Springer-Verlag.
- Day, R.A. (1998). *How to Write and Publish a Scientific Paper*, 5th edn. Westport CT: Oryx Press.
- Fisher, B.A. & Zigmond, M.J. Attending professional meetings successfully, University of Pittsburgh, http://www.survival.pitt.edu/library/documents/Attending%20Professional%20Mgt.pdf.
- Germano, W. (2001). *Getting it Published, A Guide for Scholars Serious about Serious Books*. Chicago: The University of Chicago Press.
- Haile, J.M. (2001). *Technical Style*. Central SC: Macatea Productions.
- Matthews. J.R., Bowen, J.M., & Matthews, R.W. (2000). *Successful Scientific Writing*, 2nd edn. Cambridge, UK: Cambridge University Press.
- Newberg, H. (2005). The woman physicist's guide to speaking. *Phys. Today*, **58**(2), 54–55.
- Vernon, B. (1993). *Communicating in Science*, 2nd edn. Cambridge, UK: Cambridge University Press.

Outreach and communication with the public:

- Gregory, J. & Miller, S. (1998). *Science in Public: Communication, Culture, and Credibility*. New York: Plenum Press.
- Irwin, I. & Wynne, B. (eds.) (1996). *Misunderstanding Science? The Public Reconstruction of Science and Technology*. Cambridge, UK: Cambridge University Press.
- Lindberg Christensen, L. (2007). *The Hands-on Guide for Science Communicators: A Step by Step Approach for Public Outreach*. New York: Springer.
- Wilson, A. (1998). *Handbook of Science Communication*. London: IOP Publishing Ltd.

The scientific career:

- Career Basics Advice and Resources for Scientists from Science Careers, a collection of articles issued by the American Association for the Advancement of Science, freely available from:
  `http://images.sciencecareers.org/images/careers_basics_book`.
- Dee, P. (2004). *Building a Successful Career in Scientific Research: A Guide for PhD Students and Postdocs*. Cambridge, UK: Cambridge University Press.
- Goldberg, J. (2000). *Careers for Scientific Types and Others with Inquiring Minds*. Chicago: NTC/Contemporary Publishing Group Co.
- Goldsmith, J.A., Komlos, J., & Gold, P.S. (2001). *The Chicago Guide to Your Academic Career*. Chicago: The University of Chicago Press.
- Rothwell, N. (2002). *Who Wants to be a Scientist?: Choosing Science as a Career*. Cambridge, UK: Cambridge University Press.
- Sindermann, C.J. (1958). *The Joy of Science*. New York: Plenum Press. A nice overview of the wide spectrum of scientific careers.

Writing proposals:

- Coley, S.M. & Scheinberg, C.A. (2000). *Proposal Writing*. Thousand Oaks, CA: Sage Publications.
- Friedland, A.J. & Folt, C.L. (2000). *Writing Successful Science Proposals*. New Haven: Yale University Press.
- Hailman, J.P. & Stier, K.B. (1997). *Planning, Proposing, and Presenting Science Effectively; A Guide for Graduate Students and Researchers in the Behavioral Sciences and Biology*. Cambridge, UK: Cambridge University Press.

- Hall, M.S. & Howlett, S. (2003). *Getting Funded: The Complete Guide to Writing Grant Proposals*, 4th edn. Portland State University: Continuing Education Press.
- Locke, L.F., Spirduso, W.W., & Silverman, S.J. (2007). *Proposals that Work; A Guide for Planning Dissertation and Grant Proposals*, 5th edn. Thousand Oaks, CA: Sage Publications.
- The Directorate for Education and Human Resources of the NSF has made a guide for proposal writing available on the following website:
      `http://www.nsf.gov/pubs/2004/nsf04016/nsf04016.pdf`.
  This guide is written for the preparation of proposals in undergraduate education, but it contains much useful advice for writing proposals in general.

Gender issues:

- Coiner, C. & George, D.H. (1998). *The Family Track, Keeping your Faculties while you Mentor, Nurture, Teach, and Serve*. Urbana: University of Illinois Press. A nice collection of essays on balancing career and family issues.
- Gray, J. (2002). *Mars and Venus in the Workplace*. New York: Harper Collins.
- Rosser, S.V. (2004). *The Science Glass Ceiling: Academic Women Scientists and the Struggle to Succeed*. New York: Routledge.
- Toth, E. (1997). *Ms. Mentor's Impeccable Advice for Women in Academia*. Philadelphia: University of Pennsylvania Press.
- *Beyond Bias and Barriers: Fulfilling the Potential of Women in Academic Science and Engineering*. (2007). Washington DC: Committee on Maximizing the Potential of Women in Academic Science and Engineering, National Academy of Sciences, National Academy of Engineering, and Institute of Medicine, The National Academies Press.

Appendix B   A sample curriculum

The material presented in this book lends itself well for a course for beginning graduate students. Such students are under time pressure when starting their research and taking courses in their chosen discipline. For this reason our best experience has been in teaching the course as a single-semester, 1-credit course, which amounts to approximately 15 classroom sessions. This appendix gives a sample curriculum aimed at offering instructors ideas for elements that could be included in such a course. The curriculum includes homework assignments that roughly follow the chapters in this book, but doing all assignments is unnecessary and might make the workload unacceptably high.[1] The suggested homework is intended to inspire instructors' ideas for helpful exercises.

Our experience is that it is best to teach the course to classes no larger than about 20 students. The material is conveyed most effectively in a discussion format rather than as a lecture that offers relatively little opportunity for student participation. Some topics covered are likely to touch on personal issues; many students find it easier to pose questions and share personal views and dilemmas in a small group.

Class 1. What the course and science are about. Since the course is most effective when there is ample discussion, it is important that students feel comfortable in the group. It is helpful for students to have the opportunity to introduce themselves in class and mention, for example, what they hope to learn in this course and any personal issues or questions they would like to share. Students benefit from an introduction in this class that conveys that the model of acquiring research skills by just working along with an adviser often is not the most efficient way to learn those skills. To this end, it is valuable to give them examples of the breadth of intellectual, practical, and personal skills needed for success in research. One aspect that many students find surprising and difficult to grasp is that doing science requires skills beyond logic alone (Section 2.3). It is useful to spend some time in

---

[1] Clearly, the content of this course does not present intellectual demands at the level of a typical disciplinary graduate-level course. Students, nevertheless, might be surprised to find, in this 1-credit course, a need to ponder with care issues they previously hadn't imagined require any attention.

class discussing this. We've found that a formal lecture on the nature of science, such as described in Chapter 2, is less effective than asking students to read that chapter and bring questions to the following class.

Homework. Find an example of a scientific discovery, big or small, where the essential step taken was not rooted in logic. *If you are in the position to do so, preferably give an example from your own personal experience. Write a brief essay of a few paragraphs and be prepared to talk informally about your example in the next class.*

Class 2. Choosing a research topic and adviser. Since many students who take the class don't have a research topic and adviser as yet, it is particularly worthwhile to cover the material in Chapters 3 and 4 early in the course, making students aware of considerations that can be helpful in choosing a research topic and adviser. It is an eye-opener for many students to learn that finding a research topic for which they have a passion is an essential ingredient in achieving a successful and fulfilling career. In their choice of adviser, many students tend to focus on the perceived scientific skills and reputation of the adviser, without knowing how properly to assess these traits. The importance of good personal rapport between adviser and graduate student is an issue worth emphasizing.

Homework. Interview four faculty members. *Typical for most students, you'll have occasion to talk with various faculty members in your quest to find an adviser and research topic. As an aid to discovering different styles of research, a running assignment for the course is for you to interview four faculty members about their career, their style of research, the way they combine professional and personal life, and career choices they have made. Look for both common themes and for differences. This exercise is most useful when the faculty members you choose have diverse backgrounds, experiences, and outlooks. Write a paper of approximately four pages about these interviews and hand it in at the end of the course.*

Class 3. Questions drive research. Perhaps surprisingly, many graduate students are unable to articulate *what the key question is* they aim to address in research on which they have already embarked. Worthy of interactive discussion is the role of questions as the driver of research. These questions fall naturally into a hierarchy, from large overriding ones that give direction to the research to more-detailed ones that drive daily research activities. That free association is a valuable tool for generating large numbers of questions comes as a surprise to many students, and we find it instructive to discuss the quote of Schiller in Section 5.1 in great detail. The questions thus generated can later be sorted or discarded. In the following homework problem, we use free association to generate questions about a general research topic that all students in the class know at least a little about.

Homework. Use free association to arrive at questions. *Do this exercise in a location where you will not be disturbed. Take a blank sheet of paper and a pencil and write down any and all questions that you can think of about the research topic you have been given. Don't "filter" your questions, but freely associate and write down everything that comes to mind. Do this for just an hour, and then stop. Make sure to leave empty space between the questions. Hand in a copy of your list of questions at the next class and retain the original.*

Class 4. Setting and working toward goals. It is useful to start this class with a discussion and sharing of experience with the previous homework prior to moving on to the chapter about goals. Most students know about goal-setting, but don't have a good idea *why* setting goals is so important or how best to go about doing this (Section 6.1). It is useful to discuss the five steps towards reaching goals (Section 6.2) in detail. In this class, the focus is on exciting the students about goal-setting, delaying discussion of the distinction between being goal-oriented and process-oriented (Section 6.4) to a later time. If students are assigned the homework suggested below, it is important first to talk about the ordering and prioritizing of the research questions, as described in Section 5.2. In this exercise we order the list of questions made in the previous homework assignment. It is useful to do this as a group exercise because it illustrates how a group can combine and organize the creative ideas from its different members. Each group takes the combined list of questions from all its members, and, following the steps outlined in Section 5.2, organizes them either on a large sheet of paper or on a poster generated by a computer.

Homework. Organize and prioritize questions. *Organize the questions generated by the group members in different themes, distinguish between major questions and more-detailed secondary ones and prioritize them. Include a box labeled "garbage can" with questions that you choose not to use. Use lines or color to indicate connections among questions so that the chart you produce begins to take on the character of a research plan. This exercise is about ordering of questions. It is not a beauty contest; therefore focus on content rather than form. Prepare to present the work of your group in the next class.*

Class 5. Present research plan. By spending an entire class on the presentation of the research plans made by the different groups, students find it interesting to see how different are the approaches taken by the various groups. This exercise can naturally lead to a discussion about planning of research. Students often are surprised and impressed by the added value of doing this homework in a group.

This can be a good starting point for a discussion about teamwork and the value of using others as a sounding board in research.

Homework. Because the previous homework is intensive, focus on your faculty interviews.

Class 6. Turning challenges into opportunities. Chapter 7 covers common challenges that researchers encounter in their work. It is instructive to discuss these challenges in class. Many beginning graduate students think their professor knows everything and does not encounter any such difficulties. To counteract this myth, it is beneficial for instructors to share with the students their own anecdotes, illustrating problems they have run into while doing science. Also, when students share their research challenges in class, they learn that they are not alone in encountering such problems. What a relief it is for many students to discover that this is a normal part of doing science!

In the class after this one we will have a scientist come visit. Students can ask any questions that come to mind. (It is a good idea for the students to know in advance who this visitor will be and have some background information about the person.)

Homework. Make a list of questions to pose to a visiting scientist. *Prepare a list of questions that you would like to ask the visitor to next week's class. These questions can be either personal or professional.*

Class 7. Asking questions of the visitor. Since different researchers have different work habits and lifestyles, and to avoid the bias of the instructor, it helps to have a scientist other than the instructor visit the class. In fact, a few classes could be devoted to different visitors. Having prepared a list of questions in the previous homework assignment, students pose these questions in this class. Students may ask whatever they like, no matter how personal the question, but the visitor of course has the option to decline to answer any given question. Having one of the visiting scientists be a woman invariably raises its own set of questions about the combination of professional and personal life.

Homework. Write an essay about your life as a scientist. *The previous class likely gave much food for thought about the life of a scientist. Write a one-page essay about your preferred life as a scientist. What would you like to achieve? What are your dreams? Where would you like to work? What balance would you like to strike between work and personal life?*

Class 8. Ethics of research and the meaning of work. One could readily teach ethics for an entire semester (indeed, such full courses would be a valuable addition to the curricula in any discipline), so discussion of this topic in class unavoidably needs to be limited. Chapter 8 gives ideas for topics that are most closely related to the daily activities of a scientist. Discussion of a few case studies is a great way to stir student interest in this important topic. The booklet *On Being a Scientist; Responsible Conduct in Research*[2] contains useful material for this class, including numerous case studies. This class also offers a good moment to discuss the underlying meaning of our work (Section 6.5) and the importance of being aware of larger potential consequences of the science that we do (Section 8.7). These topics can be combined in a natural way.

Homework. Write your own ethics statement. *Chapter 8 starts with an example of an ethics statement. Write your own ethics statement in such a way that this statement reflects your values as a scientist and any higher purpose to which you might aspire.*

Class 9. Visit to the library. Many students think of the library as a repository of books and are unaware of the help that the staff and facilities of the library can provide. In this class we visit the library, and one of the librarians gives an overview of different types of literature searches, the different databases that are available, and the help in accessing information that the library staff can offer. Library staff usually are delighted to assist in such a class.

The next exercise forces students to start using a database of references, as described in Section 9.3.

Homework. Create a database of references. *For the remaining duration of the course, create a database of the articles and books that you have read for your own research. First choose the database system that you want to use for archiving references. Ask your adviser and fellow students for their recommendations. If this does not help you, use* zotero, *which can be downloaded for free from* www.zotero.org. *Add to your database any relevant paper or book you envision that you might ever want to cite or read in the future. At the end of the course, provide a printout of a representative part of your database in a readable form that is ordered by the keywords that you have used in the database.*

Class 10. Oral communication. We've found it less effective to tell students how to give oral presentations than to demonstrate the difference between a good and bad presentation. We show them in class two versions of a five-minute presentation. In one version (without alerting the students in advance) we do everything deliberately wrong, while the second version is as good as we can make it. The slides in

[2] http://www.nap.edu/html/obas.

Fig. 10.1 are, in fact, taken from the presentations that we use in class. After these presentations, we ask the students to articulate why the second presentation was better than the first one. It is interesting that, no matter how clownish and ridiculous we make the first presentation, we always get the same response. Students at first don't believe that speakers sometimes (in fact, too often) actually behave in such an unprofessional way, and then, over the following weeks, they express amazement at how many of the mistakes in the "bad presentation" they witness in listening to actual presentations that are meant to be professional.

Homework. Make a list of good and bad habits of speakers. *Compile a list of habits of speakers that you find pleasing and productive, and a list of habits that are ineffective and irritating. Write a plan with at least five action items for how you will train yourself to acquire the good habits and avoid the bad ones.*

Class 11. Writing and publishing a paper. Most junior students have no idea as yet about how the publication process works and how to choose a journal for publication. The considerations and practices described in Chapter 11 are new to many students. When presenting this material, anecdotes can be useful in illustrating salient points. Some students in the class might be in a position to share some of their experiences in publishing papers.

The following exercise serves to introduce students to the important journals, *Science* and *Nature*.

Homework. Find and discuss a paper published in *Science* or *Nature*. *Find out how to access these two journals electronically and browse some recent issues. From either of these journals, select a paper that you found truly interesting or exciting. Hand in a copy of the chosen paper along with a brief description of why this paper appealed to you.*

Class 12. Time management. "Not having enough time" is a disease of our times, and graduate students struggle as much with managing their time as do other professionals. The material of Chapter 12 forms a good basis for a discussion about this topic. Figure 12.1 is an eye-opener to many students. Be aware that many students spend an inordinate amount of time dealing with email and surfing the internet. Advice and tips on how to use these tools effectively without being the slave of the tool are badly needed.

Homework. Analyze how you spend time in the coming week. *Throughout the coming week write down your activities and the amount of time you spend on each of them. Place each activity in the appropriate quadrant of Fig. 12.1, and annotate with the amount of time spent on the activities in each of the quadrants.*

Make a decision as to whether or not you want to change the way you spend your time, and, if you perceive the need to change, formulate five steps for change that you can take over the coming month. Hand in an overview of the time spent on activities in the four quadrants, a brief description of what you want to change, and the five steps that you will take to make those changes.

Class 13. Writing proposals. Most students are keenly aware of the need to write proposals, but don't know how funding agencies work and what are the elements of successful proposals. The material of Chapter 13 is a good starting point for a discussion. It can be illustrative for students to be given a copy of a proposal that was funded and use it as an example to structure the discussion.

Homework. Find a funding agency and program for your research. *Go to the websites of funding agencies and find a program that is appropriate for your research project. Write a memo to your adviser with a recommendation for a program and funding agencies that match your research, and outline why you think this is a good match. Hand in a copy of the memo.*

Class 14. The scientific career and applying for a job. Graduate students often are apprehensive about the competitiveness of the scientific career. It is important to emphasize that, while competition indeed does exist while pursuing a career in science, the rewards of the pursuit, particularly when done in cooperation with others, greatly outweigh drawbacks. Sketch in class some examples of scientific careers along with the challenges and rewards that come with these different careers. A great way to cover this topic is to invite to class a panel of scientists who have followed different career paths, but this might not be easy to realize.

Homework. Think up questions and discussion items for the next, final class.

Class 15. Closing session. The above-suggested curriculum covers much ground, so it can be helpful for students to have ample opportunity to raise any questions, concerns, or comments they have, and devote the final class completely to this purpose. Many junior graduate students are apprehensive about the road ahead. The final session can be given a light note by doing something fun, for example, by having the students out for lunch or drinks together.

Comment for instructors. As an instructor you will find it a rewarding experience to interact with students in a course similar to the one outlined here. A side effect we've encountered is that students who have taken the course have generally shown extensive willingness to come to us, seeking advice. Not unlike graduate students everywhere, a significant number of these students had experienced distress in one aspect or another of their graduate career. It can happen that, when graduate students encounter problems, their adviser might not offer the help and support that they need, especially when the adviser is part of the problem. With other faculty members being busy, students often feel there is no one else in the faculty to talk with. This leaves them without the counsel that they need at such critical moments. Students might see the instructor for this course as the natural person with whom to talk. We have found these discussions useful and rewarding for us personally. More important and more generally, instructors do students a great favor by offering them the informal advice and support that they need.

Appendix C   The Refer and BibTeX format for a database of references

In this appendix we give an overview of two common formats for a database of references. The first is the Refer format, the second is the BibTeX format. The Refer format can, for example, be used with the EndNote program for managing the database. As an example of an entry for a publication, consider the following reference.

Jones, H. & Jackson, R. (2001). Nuclear fusion in frog cells, *J. Improbable Results*, **24**, 2245–2298.

In the Refer format this reference would be entered as

```
%A Jones, H.
%A Jackson, R.
%D 2001
%T Nuclear fusion in frog cells
%J J. Improbable Results
%V 24
%P 2245-2298
%K frogs, fusion, weird paper
```

This information can be stored as a plain text file. This means that information can be entered with any text editor, but dedicated database programs such as EndNote allow the user to enter information interactively.

The structure of the format is simple; each line starts with %something. The key after the %-sign indicates the field in which the information of that line fits. For example, when a line starts with %A, this indicates that the following information gives the name of an author. For more than one author, the author line is repeated for every individual author. The field starting with %V gives the volume of a journal or book series. The line %K is for keywords. It is useful to enter well-chosen keywords with each reference because this makes it possible to select all references that are relevant for a certain topic efficiently.

The example given above is relevant for a journal article. For a book, or for a chapter in a book, other fields can be used as well. Here is a more complete list of entries that can be used in the Refer format.

%A Author's name

%T Title of the article or book

%S Title of a book series

%J Name of journal containing article

%N Number with volume

%B Title of book containing article

%E Editor of book

%P Page numbers

%I Issuer (this is the publisher)

%C City where the publisher is based

%D Date of publication. In practice, this is the year of publcation

%K Keywords

%X Abstract

The Refer format supports a larger amount of entries than are shown here, but the entries shown above are the ones that are most useful for a personal database. As shown in the example of the reference of Jones and Jackson, it is unnecessary to enter all fields shown above. The fields to be used depend on the type of reference. For a journal article, the entry %J with the journal name is relevant, but this entry is irrelevant for a book.

Normally, the pages in a volume of a journal start with 1 and are numbered sequentially in subsequent issues of that journal in the same volume. In that case, the volume number and the page number uniquely define the location of an article. However, in some journals all the issues within one volume start with page 1. In that case the volume and page number do not describe the location of an article uniquely, as a paper might have appeared in any of the issues of the journal within that volume. In that case the entry %N is used to denote the issue within that journal.

For books, it is customary to give the name of the publisher and the city where the book is published. It frequently happens that scientists contribute a chapter to a book. Then, it is customary to give the title of the book (which is different from the title of the chapter) as well as the names of the editors of the book. In the Refer format, the fields %B and %E can be used for this, respectively. The entry %E should be repeated for each editor, just as it is done for multi-author papers.

An alternative to the Refer format is the BibTeX format. This is a format for a database of references that is integrated with the word-processing program LaTeX. In BibTeX format, the article of Jones and Jackson would be entered as

```
@article = {Jones01Nuclear,
author = {Jones, H. and Jackson, R.},
title = {Nuclear fusion in frog cells},
journal = {J. Improbable Results},
volume = 24,
pages = {2245-2298},
date = 2001}
```

In BibTeX, the references are stored as a plain text file so that the database can be edited with any text editor. Unlike in the Refer format, the BibTeX format requires that a category of each reference is specified. In the example above, that is done with the @article statement. Alternative categories include @book, @incollection (for a chapter in a book), and @phdthesis. The number of possible entry types is large. A useful overview of the use of BibTeX is given by Kopka and Daly (1993).

The word Jones01Nuclear indicates a key that identifies this article. In LaTeX, a citation to the paper of Jones and Jackson can simply be made by inserting the command \cite{Jones01Nuclear} at the place in the text where the citation is made. LaTeX has separate commands for automatically creating the bibliography into a manuscript.

In BibTeX every author is entered with the last name first, followed by a comma and the initials, as shown in the example above. If there are multiple authors, these are separated with the word "and," regardless of the number of authors. The other fields in the example above are probably self-explanatory. Curly brackets are used when an entry contains blank space, commas, or other signs that can act as a delimiter. When in doubt, it is a good idea to enclose items in curly brackets. There are many other possible data fields in addition to the ones shown in the example above. These include editor, booktitle, chapter, and many others, as described in much more detail by Kopka and Daly (1993). One caveat must be made with LaTeX (and BibTeX): this word-processing package is finicky to errors in its input, especially for missing curly brackets. It can be helpful to use an editor that uses a color code to indicate missing brackets.

References

Ashby, N. (2002). Relativity and the global positioning system. *Phys. Today*, **55**(5), 41–47.

Barish, B.C. & Weiss, R. (1999). LIGO and the detection of gravitational waves. *Phys. Today*, **52**(10), 44–50.

Berry, W. (1992). *Sex, Economy, Freedom and Community*. New York: Pantheon Books.

Bronowski, J. (1956). *Science and Human Values*. New York: Julian Messner Inc.

Bronowski, J. (1973). *The Ascent of Man*. Boston: Little Brown and Co.

Bukowski, R., Szalewicz, K., Groenenboom, G.C., & van der Avoird, A. (2007). Predictions of the properties of water from first principles. *Science*, **315**, 1249–1252.

Burns, R.A. (1985). Information impact and factors affecting recall. Paper presented at the 7th Annual National Conference on Teaching Excellence and Conference of Administrators (Austin, TX), May 22–25.

Chittister, J. (1991). *Wisdom Distilled from the Daily: Living the Rule of St. Benedict Today*. San Francisco: Harper.

Cipra, B. (2000). *Misteaks . . . and How to Find Them Before the Teacher Does*, 3rd edn, Natck, MA: Peters, A.K.

Coburn, A. & Spence, R. (1992). *Earthquake Protection.* Chichester, UK: John Wiley & Sons.

Committee on Science, Engineering, and Public Policies. (1995). *On Being a Scientist: Responsible Conduct in Research.* Washington, DC: National Academy Press. `http://www.nap.edu/html/obas`.

Covey, S.R. (1990). *The 7 Habits of Highly Effective People.* New York: Fireside Books.

Cowan, C.L., Jr, Reines, F., Harrison, F.B., Kruse, H.W., & McGuire, A.D. (1956). Detection of the free neutrino: a confirmation, *Science*, **124**, 103.

Craig, C.R., Cather, A., & Culberson, J. (2003). An ethical affirmation for scientists, *Science*, **299**, 1982.

Csikszentmihalyi, M. (1990). *Flow: The Psychology of Optimal Experience.* New York, NY: Harper and Row.

Emerson, R.W. (1951). *Essays,* originally published in 1841; *reprinted in Emerson's Essays,* New York: Harper & Row Publishers.

Emery, M. (2008). The fire service pyramid of success. *Firehouse Magazine,* 84–88, April.

Epple, A. (1997). *Organizing Scientific Meetings.* Cambridge, UK: Cambridge University Press.

Feynman, R. (1998). *The Meaning of it All; Thoughts of a Citizen Scientist.* Reading MA: Addison-Wesley.

Freud, S. (1994). *The Interpretation of Dreams,* originally published in 1899; *reprinted* New York: Barnes and Noble.

Friedland, A.J. & Folt, C.L. (2000). *Writing Successful Science Proposals.* New Haven, CT: Yale University Press.

Friedman, T.L. (2007). *The World is Flat 3.0: A Brief History of the Twenty-first Century.* New York: Picador.

Goldstein, I.F. & Goldstein, M. (1984). *The Experience of Science,* New York: Plenum Press.

Goswani, A. (1995). *The Self-aware Universe.* New York: Penguin Putnam Inc.

Griffin, E. (2002). Descending the career ladder. *Astron. Geophys.,* **43**(6), 17–18.

Hadamard, J. (1954). *The Psychology of Invention in the Mathematical Field.* New York: Dover Publications.

Hannan, P.J. (2006). *Serendipity, Luck and Wisdom in Research.* Universe.

Huberman, E. & Huberman, E. eds, (1971). *50 Great Essays.* New York: Bantam Books.

Joint project of The American Council on Education, The American Association of University Professors, and United Educators Insurance Risk Retention Group. (2000). *Good Practice in Tenure Evaluation.* Available freely from: www.acenet.edu/bookstore/pdf/tenure-evaluation.pdf.

Jonas, H. (1982). Technology as a subject for ethics, *Social Res.,* **49**, 891–898.

Killeffer, D.H. (1969). *How Did You Think of That?* Garden City: Doubleday & Company.

Kirby, K. & Houle, F.A. (2004). Ethics and welfare of the physics profession. *Phys. Today,* **57**(11), 42–46.

Koestler, A. (1989). *The Act of Creation.* London: Arkana Penguin Books.

Kopka, H. & Daly, P.W. (1993). *A Guide to LaTeX.* Wokingham, UK: Addison-Wesley.

Kuhn, T.S. (1962). *The Structure of Scientific Revolutions.* Chicago: The University of Chicago Press.

Laughlin, R.B. (2002). Truth, ownership, and scientific tradition. *Phys. Today*, **55**(12), 10–11.

Lelieveldt, H. (2001). *Promoveren, een Wegwijzer voor de Beginnend Wetenschapper*. [*Graduate Studies: A Roadmap for Beginning Scientists*] Amsterdam: Aksant.

Maddox, J., Randi, J., & Stewart, W.W. (1988). "High-dilution" experiments: a delusion, *Nature*, **334**, 287–290.

Matthews, J.R., Bowen, J.M., & Matthews, R.W., (2000). *Successful Scientific Writing*, 2nd edn. Cambridge, UK: Cambridge University Press.

Moore, J. (1993). *Science as a Way of Knowing: The Foundations of Modern Biology*. Cambridge, MA: Harvard University Press.

Moore, R. (1966). *Niels Bohr: The Man, His Science, and the World They Changed*. New York: Random House.

Murray, W.H. (1951). *The Scottish Himalayan Expedition*. London: J.M. Dent & Sons Ltd.

Murray-Rust, P. (2008). Chemistry for everyone. *Nature*, **451**, 648–651.

Myers, T.M. (2001). A student is not an input. *New York Times*, March 26.

Oxman, A.D., Chalmers, I., & Liberati, A. (2004). A field guide to experts. *British Medical J.*, **329**, 1460–1462. Also available from http://www.bmj.com/cgi/content/full/329/7480/1460.

Pais, A. (1986). *Inward Bound*. Oxford, UK: Oxford University Press.

Parkinson, C.N. (1958). *Parkinson's Law: The Pursuit of Progress*. London: John Murray.

Pausch, R. (2008). *The Last Lecture*. New York: Hyperion.

Pigg, K.B. & DeVore, M.L. (2001). Paleobotany. *Geotimes*, **46**(7), 35–36.

Popper, K. (1965). *The Logic of Scientific Discovery*. New York: Harper Torchbooks. (1st edn, 1934).

Preston, A.E. (2006). *Leaving Science: Occupational Exit from Scientific Careers*. Burlington, MA: Elsevier Academic Press.

Rhodes, R. (1995). *Dark Sun: the Making of the Hydrogen Bomb*. New York: Simon & Schuster.

Robbins, A. (1997). *Unlimited Power*. New York: Fireside Books.

Robbins-Roth, C., ed. (2006). *Alternative Careers in Science: Leaving the Ivory Tower*, 2nd edn. Burlington, MA: Elsevier Academic Press.

Roberts, R. (1989). *Accidental Discoveries in Science*. New York, NY: Wiley.

Saint-Exupéry, A. (1931). *Vol de Nuit*, Paris: Gallimard.

Scales, J. & Snieder, R. (1999). Computers and creativity. *Geophysics*, **64**, 1347–1348.

Simon, A., (1998). *North to the Night*. New York: Broadway Books.

Sindermann, C.J. (1958). *The Joy of Science*. New York: Plenum Press.

Snieder, R. (2000). The tube worm turns. *Nature*, **406**, 939.

Song, M., Halsey, V., & Burress, T. (2007). *The Hamster Revolution*. San Francisco: Berrett-Koehler Publishers Inc.

Speelman, H. (1998). Policy plan 1999–2002, Netherlands Institute of Applied Geoscience TNO: National Geological Survey.

Stenger, V. (2007). *God: The Failed Hypothesis*. Amherst, New York: Prometheus Books.

Strunk, W. Jr, & White, E.B. (2000). *The Elements of Style*, 4th edn. Boston: Allyn and Bacon.

Washburn, J., University, Inc. (2006). *The Corporate Corruption of American Higher Education*. New York, NY: Basic Books.

Weglein A. (2003). CSEG interviews Art Weglein, *The Leading Edge*, **22**, 976–982.

Whitehead, A. N. (1967). *Science and the Modern World*. New York: The Free Press, Macmillan Publishing.

Wigner, E.P. (1960). The unreasonable effectiveness of mathematics in the natural sciences, *Commun. Pure Appl. Math.*, **13**, 222–236.

Zee, A., (2003). *Quantum Field Theory in a Nutshell*. Princeton: Princeton University Press.

About the authors

Roel Snieder holds the Keck Foundation Endowed Chair of Basic Exploration Science at the Colorado School of Mines. In 1984 he received a Master's degree in Geophysical Fluid Dynamics from Princeton University, and in 1987 a Ph.D. in seismology from Utrecht University. For this work he received the Vening Meinesz Award from the Netherlands Organization for Scientific Research. In 1988 he worked as a postdoctoral fellow in the "Equipe de Tomographie Géophysique" at the Université Paris VI, and was appointed in 1989 as Associate Professor at Utrecht University. In 1993 he was promoted to Full Professor of seismology at Utrecht University, where from 1997–2000 he served as Dean of the Faculty of Earth Sciences and spearheaded the integration of the research of the department and the Netherlands Institute of Applied Geoscience TNO. Roel served on the editorial boards of *Geophysical Journal International, Inverse Problems*, and *Reviews of Geophysics*. In 2000 he was elected a Fellow of the American Geophysical Union for important contributions to geophysical inverse theory, seismic tomography, and the theory of surface waves. He is author of the textbook *A Guided Tour of Mathematical Methods for the Physical Sciences*, published by Cambridge University Press. Since 2003 he has been a member of the Earth Science Council of the US Department of Energy. In 2008 Roel worked for the Global Climate and Energy Project at Stanford University on outreach and education on global energy. Roel has been a volunteer firefighter since 2000 and currently is a Captain with Genesee Fire Rescue. For more information, visit his website http://www.mines.edu/~rsnieder.

 **Ken Larner** received the degree of geophysical engineer from the Colorado School of Mines (CSM) in 1960 and a Ph.D. in geophysics from MIT in 1970, after serving in Vietnam in the U.S. Army. He joined Western Geophysical Company as a senior research geophysicist in 1970 and, in 1988, left his position as Western's vice president for geophysical research to become the Charles Henry Green Professor of Exploration Geophysics at CSM. He received best presentation and best paper awards from the Society of Exploration Geophysicists (SEG), the Canadian SEG, Offshore Technology Conference, Australian Petroleum Exploration Association, and the International Symposium on Exploration Geophysicists in Beijing. He is a past 1st Vice President and Honorary Member of the Geophysical Society of Houston, He received the 1988 Conrad Schlumberger Award of the European Association of Exploration Geophysicists and, in 1990, the Medal of the Society of Venezuelan Geophysicists. In 1996, he received the SEG's highest award, the Maurice Ewing Medal, and is an Honorary Member of the SEG. That same year, he was inducted as a Foreign Fellow of the Russian Academy of Natural Sciences. He was Spring 1988 SEG Distinguished Lecturer and a 2000–2001 Distinguished Lecturer for the Society of Petroleum Engineers. He served as 1st Vice President of the SEG (1979–80) and SEG president in 1988–89 and, from 1996 through 2004, as Director of the Center for Wave Phenomena at CSM (a research consortium sponsored by 27 international companies in the oil exploration industry).

Index